Scottish Placenames

Scottish Placenames

David Ross

BIRLINN

This edition published in 2025 by
Birlinn Limited
West Newington House
10 Newington Road
Edinburgh
EH9 1QS

www.birlinn.co.uk

ISBN 978 1 78027 924 4

British Library Cataloguing-in-Publication Data
A catalogue record for this book is available
from the British Library.

Typeset by Mark Blackadder

MIX
Paper | Supporting
responsible forestry
FSC® C018072

Printed and bound by Clays Ltd, Elcograf, S.p.A.

What's in a Placename?

A surprising amount of history is packed into Scotland's placenames. Names always have a meaning, and can tell us about who lived there and when, what they thought was important about it, and what language they spoke. They can give valuable hints as to their inhabitants' beliefs and and daily activities. Placenames can also shed light on large-scale happenings like the Viking invasions, or how the use of Scots spread across across a largely Gaelic-speaking country. Many placenames have also become surnames, sometimes maintaining an ancient form.

How to use this book

Each entry starts with a present-day name, and its location in a region or district. The original meaning is then given, followed by the original forms of the words that make it up, and their language, with any necessary explanation. Sometimes there is more than one possible derivation. Finally the earliest record of the name is given, often with later forms to show how the present name has evolved.

Useful things to know

Language: Some placenames began in modern Scots or English, but most are older. The majority are from Gaelic, but many are from Brittonic, some from Pictish, some from Old Norse and the Anglian form of Early English. A few reveal a Latin source, and a handful are borrowed from French or German names.

Northern Brittonic: Brittonic was the language spoken by most of the population of Britain south of the Forth-Clyde line at the time of the Roman invasions of the first century CE. It was a 'Celtic' language, related to the Gaulish speech of the continent. The form spoken in what is now Scotland is known as Northern Brittonic but for convenience in this book is simply 'Brittonic'.

Pictish: The language spoken north of the Forth-Clyde line, including Orkney and Shetland. It was related to Brittonic but with variations, some of which appear in placenames.

Anglian: The Early English (Germanic) speech of continental settlers who established a colony at Bamburgh in Northumberland in the mid-6th century.

Scots: The speech that evolved from Anglian, influenced by Gaelic and Brittonic as it spread northwards into Lothian.

Gaelic: Also a 'Celtic' language. Its early form developed in Ireland with settlers from Europe. Scottish Gaelic dates from the late 5th century, when Gaelic speakers moved into Argyll. 'Gaelic' in this book refers to the Scottish form, and Irish Gaelic is noted as such.

Old Norse: Early forms of Norwegian and Danish, spoken by the Vikings who colonised the northern and western islands and part of the mainland from 790CE.

Latin: Spoken in Scotland only by Roman invaders of the 1st and 2nd centuries CE, but has left some traces in placenames.

Continental Celtic: Certain names, especially of rivers, have been traced back to languages spoken by the European ancestors of Britons and Gaels. No documents survive, and such names have been traced back to basic forms known as root-words. They are identified with an asterisk, e.g. *ona*, to show they have been established by modern linguists.

Language movement: Speech has always had a political and societal dimension. Around 100CE all the inhabitants, organised into tribal groups, spoke Brittonic or the related Pictish. By 547, Anglian speakers from the South were in control of Lothian. The arrival of Gaelic-speaking Scots, the introduction of Christianity from Ireland, and rivalry between Scots and Picts, led to the moment in 843CE when Picts and Scots were forcibly united under a Gaelic-speaking king. This led to the gradual decline and extinction of Pictish on the mainland.

Attacks and settling by the Vikings from 790 onwards also brought the end of Pictish in the Northern Isles. In the Hebrides, Old Norse and Gaelic co-existed, but with the assimilation of the Norse speakers into the Scottish kingdom, Gaelic survived and Old Norse died out. A form of Old Norse called Norn lingered on in the Northern Isles until around 1850.

Meanwhile Brittonic was assailed from all directions: Anglian from the East, Old Norse from West and South, Gaelic from the North. By the 9th century their once-strong kingdom of Strathclyde was ruled by Gaelic kings and would be be assimilated into the new realm of Alba, soon to take the name of Scotland. In 889 the chiefs of Strathclyde and many followers removed to join other Brittonic speakers in Wales. In 1034 Strathclyde ceased to be even a sub-kingdom. With the gradual establishment of a unified state, Gaelic for a time became predominant. But Anglian, with its southern connection, survived in the rich Lothian area, and the fortress of Din Eidynn came to replace Dunfermline as capital. Anglian was essential for diplomacy, and from around 1100, evolving towards Scots, with elements both from Brittonic and Gaelic, it gradually became the everyday language south of the Highlands. The Highlands and Western Isles became the *Gaidhealtachd,* with Gaelic surviving as the hearth tongue into the 19th century. The rise of literacy, the centralising of government, and modern communications ensured that by around 1900 there were no monoglot Gaelic speakers left. Gaelic is now a second language of choice for those who wish to explore its past and assure its future.

Many old names stayed in use even when their language was no longer spoken. Very often they underwent alteration or addition. Thus with Old Norse *dalr* 'valley' forgotten, Glen was added, as in Glen Borrodale. All this explains why in some areas a name can be either Brittonic, Old Norse, Gaelic or Scots, or a hybrid mixture.

Older Sources: The sparse earliest records are in Latin. The names 'Albion' meaning the entire island, also the nearby 'Ierne', were noted by a Carthaginian voyager, Himilco, around the end of the 5th century BCE. Around 320 BCE a Greek navigator, Pytheas, noted the names 'Orca' (Orkney), and 'Hebudes'. The Roman historian Tacitus, accompanying the army in 80/85CE, mentions 'Clota' (Clyde). Around 150 CE the Greek geographer Ptolemy of Alexandria recorded some 60 location names. From the 6th century onwards, placenames can occasionally be found in surviving church documents. The Old Welsh *Gododdin* poem, on the deeds of the Lothian-based Votadini tribe, gives Din Eidynn. The 7th century Italian *Ravenna Cosmography* lists a few names. Bede's *Ecclesiastical History of the English People* (731) gives more. From the 12th century onwards, the recording of landholdings gives many, including Gaelic names in the *Book of Deer*. The *Pictish Chronicle*, a 14th century transcription of 10th century manuscripts, gives a few placenames. From the 17th century, maps provide much useful information: maps of Scotland made by Timothy Pont and revised by Robert Gordon are included in Blaeu's *Theatrum Orbis Terrarum*, published in Amsterdam in 1654.

Note on translations: In Gaelic and Brittonic, many words have multiple meanings. Also, all nouns are either masculine or feminine. Gaelic *càrr* (fem.) can mean 'itch' or 'rocky shelf'; *càrr* (masc.) can mean 'bog' 'or 'shingle on hilltop'. The context usually makes the meaning clear.

This book is based on the author's *Scottish Placenames* (2001), with many updatings and corrections.

Gaelic pronunciation: some very basic hints

accents **à, ì,** indicate a long vowel

ao is sounded **uh,** with a hint of **y**

bh at the start of a word as **v,** elsewhere as **v** or **w**

d before **e** or **i** is sounded **dj-**

dh at the end of a word is not pronounced

gh is sounded as a very soft **g**

l and **r** are given full value, almost as if followed
by a short **i**

mh is sounded as **v**

s preceded or followed by **e** or **i** is sounded **sh**

sg is sounded as **sk**

t before **i** or **e** is prounced **tch-**

th at the start of a word is sounded **h**. After or between
vowels it is not sounded

The Scottish Place-Name Society (Comann Ainmean-
Àite na h-Alba) welcomes those with an interest in the
history and meaning of Scotland's placenames.
    See www.spns.org.uk.
    This book is an independent publication.

A

Aber- This prefix normally indicates a place where two rivers join, or a river enters the sea or a loch. It existed in both Brittonic and Pictish. The equivalent Gaelic term is *Inbhir*, scotticised to Inver. Aber- names are most frequently found on the eastern side of the country. North of Inverness and on the west coast they are extremely rare. Many Aber- names have Gaelic forms following the Pictish prefix, suggesting that the prefix was retained for a reason (see also Pit- names).

Abercairney (Perth & Kinross) 'Confluence by the thicket' or 'cairns'. *Aber* (Pictish) 'confluence', *cardden* 'thicket'; with *-ach* (Gaelic suffix) indicating 'place'; or alternatively *càirneach* (Gaelic) 'place of cairns or rough rocks'. Recorded in 1218 as Abercarnich.

Aberchirder (Aberdeenshire) 'Confluence of the dark water'. *Aber* (Pictish) 'confluence'; *ciar* (Gaelic) 'dark', *dobhar* 'waters'. Noted c. 1212 as Aberkerdouer.

Abercorn (West Lothian) Perhaps 'horned confluence'. *Aber* (Pictish-Brittonic) 'confluence'; *corniog* (Brittonic) 'horned': a reference to the 'horn' between two joining streams (one still called the Cornie Burn). Bede notes it (731) as Aebbercurnig.

Aberdeen 'Mouth of the River Don'. *Aber* (Pictish) 'river-mouth'; the name was recorded as Aberdon in the 12th century, referring to the original settlement now known as Old Aberdeen. It may first have meant 'confluence of the Den (Burn)' which joins the Don. Noted as Aberdoen 1178; Aberden 1214. See also rivers Dee and Don.

Aberdour (Fife; Aberdeenshire) 'River-mouth'. *Aber* (Pictish) 'river-mouth', 'confluence'; *dobhar* (Gaelic) 'waters'. The Fife name is found in 1126 as Abirdaur. Aberdour parish in Buchan is noted in the *Book of Deer* c. 1100, as Abbordoboir.

Aberfeldy (Perth & Kinross) 'The confluence of Pallidius or Paldoc'. *Aber* (Pictish) 'confluence'; *phellaidh* (Old Gaelic) refers to St Paldoc, missionary to the Picts in the 5th century, though also the name of a local *uruisg* 'water sprite'.

Aberfoyle (Stirling) 'Confluence of the pool'. *Aber* (Pictish) 'confluence'; *phuill* (Gaelic) 'of a pool'. Found in the 11th century as Eperpuill; 1481 as Abirfull.

Aberlady (East Lothian) 'Swampy estuary'. *Aber* (Brittonic) 'river-mouth'; Gaelic *lobh*, 'putrefy', although the stream once bore the name Peffer, 'shining' (see Strathpeffer). Noted as Aberleuedi, 1214.

Aberlour (Moray) 'Loud confluence'. *Aber* (Pictish) 'confluence', *labhar* (Gaelic) 'loud'. Its later name of Charleston of Aberlour comes from that of Charles Grant who developed the 'new' village in 1812.

Abernethy (Perth & Kinross; Highland) 'Confluence of the bright river'. *Aber* (Pictish) 'confluence'; the second part may be from *na eitighich* (Gaelic) 'of the gullet', but it is likely to be pre-Gaelic. The Brittonic river-name *Nedd* stems from a conjectured Celtic root-word *nido, indicating 'gleaming', and is also the root of Nidd and Nith. The Tayside name was noted as Apurnethige in the *Pictish Chronicle*, c. 970.

Abington (South Lanarkshire) 'Albin's village', found as Albintoune, 1459. *Ael-wine* (Old English personal name, *tūn* 'enclosure', 'settlement'.

Aboyne (Aberdeenshire) 'White cow ford'. *Ath* (Gaelic) 'ford', *bó* 'cow', *fhionn* 'white'. Older forms include Obyne, 1260; modern Gaelic is A-bèine.

Achanalt (Highland) 'Field by the river'. *Achadh* (Gaelic) 'field', *an* 'by the', *uillt* 'of the stream'.

Acharacle (Argyll & Bute) 'Torquil's ford'. *Ath* (Gaelic) 'ford'; *Torcuil* (Gaelic–Norse proper name) 'Torquil', from Thorketil.

Achiltibuie (Highland) May be 'Yellow height', Pictish-Gaelic hybrid gaelicised as *Achillidh* 'height', from Pictish **ochel* 'high' with *buidhe* (Gaelic) 'yellow'. The explanation as 'field of the yellow(-haired) lad', from *Achadh-a-gille-buidhe*; with *gille* (Gaelic) is less likely.

Achnacloich (Highland; Argyll & Bute) 'Field of stones'. *Achadh* (Gaelic) 'field', *na* 'of', and *cloich* 'stones'.

Achnahannet (Highland) 'Field of a patron saint's church', *Achadh* (Gaelic) 'field', *na h-* 'of the', *annait* 'church of a patron saint', 'church with relics'. *See* Annat.

Achnasheen (Highland) 'Field of the storms'. *Achadh* (Gaelic) 'field', *na* 'of the', *sian* 'storm'.

Achnashellach (Highland) 'Field of willows'. *Achadh* (Gaelic) 'field', *na* 'of', *seileach* 'willow trees'. Found as Auchnashellicht, 1543.

Achray, River and **Loch** (Stirling) Possibly 'ford of shaking'. *Ath* (Gaelic) 'ford', *chrathaidh* 'shaking' (*see* Crathie). The Gaelic name is Loch Ath-Chrathaigh.

Adder, River (Borders) 'Stream'. 'Black' and 'White' are marks of identity rather than description. Adder is one of the numerous river-words going back to continental Celtic; conjectured as **adara*, indicating 'flowing water'. The immediate derivation is probably from Old English *aedre* 'stream'. Found as Blacedre, c. 1098. See Edrom.

Addiewell (West Lothian) 'Adam's Well'. *Addie* (Scots) diminutive form of 'Adam'.

Ae, River and **Forest** (Dumfries & Galloway) 'Water'. A river-name, from Old Norse *aa* 'water'. *See* River E.

Affric, River, Glen, Loch (Highland) Possibly 'speckled ford', or 'ford of the trout', or 'of the boar'. *Ath* (Gaelic) 'ford', *breac* 'speckled,' 'trout'; or *bhraich* 'boar'. Noted as Auffrik, 1538.

Afton (East Ayrshire) 'Brown stream'. *Abh* (Gaelic) 'stream'; *donn* (Gaelic) 'brown'.

Ailort, Loch (Highland) Perhaps from *él* (Old Norse) 'snow shower', and *fjordr* 'sea inlet', 'fiord', with *loch* (Gaelic) 'lake', 'loch'.

Ailsa Craig (South Ayrshire) Perhaps Old Norse in origin, 'Ael's isle', from *Ael* (Old Norse personal name) and *ey* 'island', with Scots 'Craig' added after the significance of the -*ey* had been lost. 'Fairy rock' has been suggested, from *Aillse* (Gaelic) 'fairy', *creag* 'rock'. Also, *ail* (Old Gaelic) 'steep rock'. The oldest Scots form is Ailsay, 1404.

Aird (Highland) 'The high ground'. *Airde* (Gaelic) 'height'; there are numerous Airds.

Airdrie (North Lanarkshire) 'High hill slope'. *Airde* (Gaelic) 'height', *righe* 'slope'. Found as Airdrie, 1584.

Airth (Stirling) 'Level green place', from *àiridh* (Gaelic), which apart from 'summer pasture' can mean 'level green place'. Or conversely *àirdh* 'height', from the Hill of Airth.

Aith (Shetland) 'Isthmus', 'neck of land'. *Eidh* (Old Norse), 'isthmus'. There are three populated Aiths in Shetland; also Aithsetter ('farm on the aith' from *saetr* 'farm').

Alba (Scotland) The Gaelic name for Scotland, originally applying to the combined kingdoms of the Picts and Scots, but nowadays referring to the whole country. In the oldest texts the name refers to the whole island of Britain, and the original root probably means 'the Earth'.

Alder, Ben (Highland) 'Mountain of falling water'. *Beinn* (Gaelic) 'mountain', *aill dobhar* respectively 'rock' and 'water'. The Alder Burn would thus appear to have given its name to the mountain. See also Ben Avon.

Alexandria (West Dunbartonshire) Named after the local member of parliament, Alexander Smollett, around 1760.

Alford (Aberdeenshire) 'High ford'; the likeliest derivation is from *àth* (Gaelic) 'ford', and *àrd* 'high'. Noted as Afford around 1200.

Aline, Loch (Highland) 'The beautiful one'. *Alainn* (Gaelic) 'beautiful', and *loch* 'lake', 'loch'.

Allan, River and Strath (Stirling; Borders) The River Allan flowing through Strathallan, and the Allan Water, a tributary of the Teviot, share a pre- or early-Celtic root *alauna*, meaning 'flowing'.

Alligin, River and **Ben** (Highland) The original name is that of the river, Ailiginn in Gaelic. Its origin is uncertain, perhaps from Gaelic *àilleag*, 'jewel'.

Alloa (Clackmannanshire) 'Rocky plain'. Derived from a compound word *ail-mhagh* (Gaelic) 'rocky plain'. Noted as Alveth, 1357.

Alloway (South Ayrshire) 'Rocky plain'. Derivation as for Alloa; noted as Auylway, c. 1340.

Almond (West Lothian; Perth & Kinross) The river-name is from pre-Celtic **Ambona*, deriving from an Indo-European root-word meaning 'water'. The Gaelic name is Abhainn Aman.

Alness (Highland) 'Stream place'. Recorded as Alenes and Alune in the 13th century, it probably has the same early or pre-Celtic river-name origin, **alauna*, as Allan. The *-ais* suffix is found as an indication of 'place' in many locations in the former Pictland. The Gaelic form is Alanais.

Alsh, Loch (Highland) Derived from *aillse* (Gaelic) 'fairy', 'spectre'; an earlier spelling is Loch Ailsh, and the Gaelic name is Loch Aillse. It suggests an archaic origin, related to belief in a water-spirit.

Altnabreac (Highland) 'Trout stream'. *Allt* (Gaelic) 'river', *na* 'of', *breac* 'trout', linked with *breac* 'speckled'.

Altnaharra (Highland) 'Walled or embanked stream'. *Allt* (Scottish Gaelic) 'stream', *na* 'of', *earbhe* 'wall'. A similar form is Altnaharrie.

Alva (Clackmannanshire) 'Rocky plain'. The derivation is the same as that of Alloa.

Alyth (Perth & Kinross) 'Steep bank or rugged place'. *Aileach* (Gaelic) 'mound' or 'bank'; alternatively, *aill* (Gaelic) 'steep rock'.

Amulree (Perth & Kinross) 'Ford of Maelrubha'. *Ath* (Gaelic) 'ford', *Maelrubha* (personal name of 7th century missionary). The Gaelic name is Ath Maol-Ruibhe.

An Teallach (Highland) 'The anvil'. *An* (Gaelic) 'the', *teallach* 'anvil'. See Challoch.

Ancrum (Borders) 'Bend on the River Ale'; older forms include Alnecrumba (12th century): Ale from a pre-Celtic form *alauna*, 'water'; *crum* from Brittonic *crwm* 'bend'.

Anderston (Glasgow) 'Andrew's farm'. The origin of the suffix is old English *tūn*, 'farmstead', but once established as Scots *-ton*, it has also been attached to modern places, like this, with no ancient links.

Angus County name generally taken as commemorating the 8th-century king of the Picts, Aonghus or Oengus (Pictish personal name), who died in 761.

Annan (Dumfries & Galloway) On the basis of the latinised form *Anava*, found in the *Ravenna Cosmography*, it is a river-name, deriving from *Anu*, the Celtic goddess of prosperity; *an* is also an obsolete Gaelic term for 'water'. Noted in 1152 as Stratanant (with Brittonic *ystrad* 'valley').

Annat (Highland; Argyll & Bute; other areas). Irish Gaelic *andóit* indicates 'church holding relics of its founder'. Annat names mostly have evidence of an ancient church or burial ground, often by a stream, suitable for the rite of baptism.

Anstruther (Fife) 'The little stream'. *An* (Gaelic) 'the', *sruthair* 'little stream'. Recorded as Anestrothir 1205, Anstrother 1231.

Antonine Wall (West Lothian to West Dunbartonshire) A Roman fortification of the late first century AD that extended from Kinneil on the Forth to Old Kilpatrick on the Clyde. It was named after the Roman emperor Antoninus Pius (AD 86–101).

Aonach Eagach (Highland 'Steep notched ridge'. *Aonach* (Gaelic) 'steep hill', normally applied to ridged mountains, *eagach* 'notched'.

Appin (Argyll & Bute) 'Abbey lands'. *Apuinn* (Gaelic) 'abbey lands'. The abbey was on Lismore.

Applecross (Highland) 'Mouth of the Crosan River'. *Aber* (Pictish) 'river-mouth', *Crosan* (Pictish river-name of uncertain derivation). The *Annals of Tighernach*, c. 731, refer to it as Aporcrosan. Apart from Lochaber, this is virtually the only Aber- name in the West Highlands.

Arbirlot (Angus) 'Confluence of the Elliot Water'. *Aber* (Pictish) 'confluence'. See Elliot.

Arbroath (Angus) 'Mouth of the turbulent stream'. Scots Aberbrothock. *Aber* (Pictish) 'river mouth'; the second element refers to the local burn, from the conjectured Pictish form **brudaca*, cognate with *brothach* (Gaelic) 'filthy', but perhaps here 'seething', 'turbulent', as in the related Gaelic *bruth* 'hot'. Found as Aberbrudoc in 1189.

Arbuthnot (Aberdeenshire) 'Confluence of the holy stream' from *Aber* (Pictish) 'confluence', *buadhnat* (Gaelic) 'little virtuous one'. Aberbothenoth, 1242.

Ardeer (North Ayrshire) 'Western headland'. *Airde* (Gaelic) 'height', 'headland', *iar* 'west'.

Ardentinny (Argyll & Bute) Although suggested as 'heights of the fox': *àrd* (Gaelic) 'high', *an t-sionnaigh* 'of the fox' the final element is more likely to be *teine* (Gaelic) 'fire', 'beacon'. See Craigentinny.

Ardersier (Highland) Possibly 'high western promontory'. *Ard* (Gaelic) 'high', *ros* 'promontory', *iar* 'west'. Noted as Ardrosser in 1227. Since 1623 it has had the alternative name of Campbelltown.

Ardgay (Highland) 'Height of the wind'. *Airde* (Gaelic) 'height', *gaoithe* 'wind'.

Ardgour (Highland) Possibly 'promontory of Gabran'. *Airde* (Gaelic) has the sense of 'promontory' as well as 'height'; *Gabran* (Gaelic personal name), but *gobhar* 'of the goats' is also possible.

Ardkinglas (Argyll & Bute) 'Height of the dog-stream'. *Airde* (Gaelic) 'height', *con* (Gaelic) 'dog,' 'wolf', *glas* 'water'. The Gaelic form is Aird-chonghlais.

Ardle, River and **Strath** (Perth & Kinross) Perhaps 'river dale'. Old Norse *aar* 'water', *dalr*, 'valley', gaelicised into Srath Ardail.

Ardler (Perth & Kinross) Perhaps 'height of the mares', from Gaelic *àird* 'height' and *làir* 'of the mares'. Older forms include Ardlair and also Ardley.

Ardlui (Argyll & Bute) 'Height of the calves'. *Ard* (Gaelic) 'high', *laoigh* 'calves'.

Ardmair (Highland) 'Finger promontory'. *Àirde* (Gaelic) 'promontory', 'height', *mheàra* (from *meur*) 'finger'; *meara* 'airy', has also been suggested.

Ardmore (Highland, Argyll & Bute) 'Big height'. *Airde* (Gaelic) 'height', *mór* 'big'. The name of numerous locations in these and other regions.

Ardnamurchan (Highland) Probably 'promontory of the otters'. *Airde* (Gaelic) 'promontory, *na* 'of the', *muir-chon* literally 'sea dogs'. Another derivation of the last two elements suggests 'piracy': *muir-chol* (Gaelic) 'sea villainy'.

Ardoch (Dumfries & Galloway; Perth & Kinross; Highland) 'High place'. *Ard* (Gaelic) 'high', *-ach* (Old Gaelic suffix) 'land'.

Ardrishaig (Argyll & Bute) 'Height of bramble-bushes by the bay'. *Airde* (Gaelic) 'height', *dris* 'brambles', 'brier', *aig* (Gaelic form of Old Norse *vik*) 'bay'.

Ardross (Highland, Fife) 'Height of the promontory'. *Airde* (Gaelic) 'height', *rois* 'of the promontory'. The name refers to the high ground between the Cromarty and Dornoch Firths, and to the rising land of the coast between St Monans and Elie, where the alternative Gaelic meaning of 'woodland' is equally possible.

Ardrossan (North Ayrshire) Perhaps 'height of the little cape'. *Airde* (Gaelic) 'height', *rois* 'cape', 'headland', *-an* 'little'. Noted in this form in 1375.

Ardtornish (Highland) 'Thori's cape'. *Airde* (Gaelic) 'promontory'; *Thori* (Old Norse personal name), *naes* 'point'. Gaelic 'aird' is a later prefix when the sense of 'ness' was lost. The English 'point' adds a further superfluous word.

Argyll (Argyll & Bute) 'District or coastland of the Gaels'. *Airer* (Gaelic) 'coastland', *Gaidheal* 'of the Gaels'. The Gaelic-speaking Scots, originating in Ireland, colonised much of the western seaboard during the 6th to 9th centuries. Recorded as Arregaithel in the *Pictish Chronicle*, c. 970. The Scots form, Argyle, was used for many centuries.

Arisaig (Highland) 'River-mouth bay'. *Ar-óss* (Old Norse) 'river mouth'; *aig* (from Old Norse *vik*) 'bay'. See also Aros.

Arkaig, Loch (Highland) Probably 'dark water'. The Gaelic form, Loch Airceig, may stem from **arc* (Celtic root form) 'dusky'; *airc* (Gaelic) 'strait' has also been suggested, though its meaning is largely figurative. The *-aig* ending here is of uncertain origin.

Arklet, Loch (Stirling) Perhaps 'dark water', of similar derivation to Arkaig, or 'steep-sloped' from *airc* (Gaelic) 'difficulty', 'strait', and *leathad* 'slope'.

Armadale (West Lothian; Highland) Named 1790 after a local landowner, lord Armadale, who took his title from Armadale on the north coast of Sutherland. The likely meaning, *Arm-r* (Old Norse) 'arm', 'arm-shaped', *dalr* 'dale', 'valley', also applies to Armadale on Skye. The Old Norse personal name *Eorm* has also been suggested for the first part.

Arnprior (Stirling) 'The Prior's land'. *Earann* (Gaelic) 'portion or share of land', *na* 'of', with English 'prior'. A hybrid Gaelic–English combination; the priory is that of Inchmahome, in the 'Lake' of Menteith.

Aros (Argyll & Bute) 'River mouth'; *á* (Old Norse) 'water', *óss* 'river-mouth'. As *óss* on its own implies the river, the reason for the prefix is unclear. Gaelic *àros* 'house' or 'palace', has been suggested; noted as dun Aros, 1410.

Arran (North Ayrshire) Most likely 'place of peaked hills'. *Aran* (Brittonic) 'height', 'peaked hill'. Alternatively it may be the same as the Irish Aran Islands where *arainn* (Irish Gaelic) 'kidney', implies an arched ridge. Arran is the oldest known form, from 1154; 13th century forms include Araane, 1251; Aran, c. 1294.

Arrochar (Argyll & Bute) Explained as a Gaelic form of *aratrum* (Latin) 'plough, indicating the area of land eight oxen could plough in a year at thirteen acres each. Gaelic *àr* can mean 'ploughing' or 'battle'; *aroch* can mean 'hamlet'. Early recordings include Arathor 1248; Arachor 1350.

Arthur, Ben (Argyll & Bute) 'Arthur's Mountain', in Gaelic Beinn Artair. The modern alternative, 'The Cobbler', known from around 1800, originally referred to the central peak, and is said to be a translation of Gaelic *an greasaiche crom*, 'the crooked shoemaker'.

Artney, Glen (Perth & Kinross) 'Pebbled'. *Artein* (Gaelic) 'pebble', presumably with reference to the valley sides or floor.

Assynt (Highland) Perhaps '(land) seen from afar'. *Asynt* (Old Norse) 'visible', referring to the area's isolated peaks as seen from out at sea in the Minch. Old Norse *àss* 'ridge' has also been put forward for the first element.

Athelstaneford (East Lothian) 'Athelstan's ford'. King Athelstan, of Mercia and Wessex, conquered Northumbria and invaded Lothian in the early 10th century. In 934 he was defeated near this village, sometimes claimed to commemorate his name. But perhaps it is Gaelic *àth-ail*, 'stone ford', with Scots -ford added when the Gaelic meanings were lost.

Atholl (Perth & Kinross) The ath- element has been connected to *ath* (Gaelic) 'next'; also to *àth* 'ford'. The second part has been derived from *Fhodla* (Irish Gaelic), a traditional name for Ireland. It could thus be either 'next or new Ireland' or 'ford of Fodla'. In the *Annals of Tighernach*, c. 739, it is noted as Athfoithle.

Attadale (Highland) 'Ata's valley'. *Dalr* (Old Norse) 'valley', prefixed by the Old Norse personal name *Ata*.

Attow, Ben (Highland) 'Long mountain'. *Beinn* (Gaelic) 'mountain', *fhada* 'long'. There is another Beinn Fhada on Mull.

Auchendinny (Midlothian) Taken as 'Field of the height or fortress'. *Achadh* (Gaelic) 'field'; *denna* (Old Irish, genitive of *dind*) 'of the height'. But the location is a steep valley and perhaps a hybrid *achadh* with Old English *denu* 'valley' is possible. Noted as Aghendini, 1335.

Auchenshuggle (Glasgow) 'Rye field': *achadh* (Gaelic) 'field'; *na* 'of the', *seagal* 'rye'.

Auchinleck (East Ayrshire) 'Field of the flat stones'. *Achadh* (Gaelic) 'field', *na* 'of the', *leac* 'flat stones', found as Auchinlec, 1239.

Auchmithie (Angus) 'Field of the herd'. *Achadh* (Gaelic) 'field', *muthaidh* 'herd'.

Auchterarder (Perth & Kinross) 'Upland of high water'. *Uachdar* (Gaelic) 'upper' (land), *àrd* 'high'; *dobhar* (Brittonic) 'water'; found as Uchterardouere, c. 1200.

Auchterless (Aberdeenshire) 'Upland of the enclosed field'. *Uachdar* (Gaelic) 'upper', *liós* 'enclosure', 'garden'. Found as Uchterless, c. 1280.

Auchtermuchty (Fife) 'Upper pig enclosure'. *Uachdar* (Gaelic) 'upper land', *muc* 'pig', *garadh* 'enclosure'. Early records show Huedirdmukedi 1250, Utermokerdy 1293, Utremukerty 1294.

Auldearn (Moray) Traditionally derived as 'river of Erin'. *Allt* (Gaelic) 'river', *Eireann* (Irish Gaelic) 'of Erin'. But Earn predates the Scots' arrival, as one of many river-names derived from a pre-Celtic root form, perhaps *ar-* or *er-*, indicating flowing water. The form Aldheren is found in 1298. *See* Earn.

Auldgirth (Dumfries & Galloway) 'Stream of the enclosure', from *allt* (Gaelic) 'stream', *gart* 'enclosure field'.

Auldhouse (South Lanarkshire) 'Stream of the ghost'. *Allt* (Gaelic) 'stream', *fhuathais* 'spectre', 'apparition'.

Aultbea (Highland) 'Stream of the birches'. *Allt* (Gaelic) 'stream', *beithe* 'birches'. See Beith.

Averon, River (Highland) Perhaps shares a name with Aveyron of southern France. Both may stem from the Celtic prefix *ab-*, *av-*, indicative of a stream; while the latter element seems cognate with that of Deveron.

Aviemore (Highland) 'Big face'. *Agaidh* (Gaelic) 'face', *mór* 'big'. The first element may refer to a hill feature; also suggested as a gaelicisation of a lost Pictish word.

Avoch (Highland) 'Place of the stream'. *Abh* (Gaelic) 'water', 'river', stemming from the same continental celtic root **ab* or **av* as Averon; *-ach* (Gaelic) a variant on *achadh*, 'field'. Noted as Auauch, around 1333; Obhach in Gaelic.

Avon, River, Loch, Ben (Highland; Moray; West Lothian) 'Stream', from *abhainn* (Gaelic) 'stream', 'river'. In Lanarkshire and Stirling it derives from Brittonic **abona*. Both stem from the conjectured Indo-European root form **-ab* or **-au*, seen also in Danube and Punjab. The mountain takes its name from the river.

Awe, River and **Loch** (Argyll & Bute, Highland) 'Water', from *àbh* (Gaelic) 'water' (see Avoch). The Argyll Loch Awe is often found in pre-20th century texts as Lochow.

Ayr (South Ayrshire) The town is named from the river: in Gaelic Inbhir-àir, 'Ayr-mouth'. Ayr has many similar forms in England and Europe (Aire, Aar, Ahr, Ohre, Ore). It has been traced to a hypothetical Celtic form, **ara, with the original sense of 'smooth-running'. Noted as Ar, 1177.

Ayre (Orkney) 'Tongue of land', from Old Norse *eyri* 'tongue', 'spit'. There are numerous ayres, joining two islands, or linking an island to the mainland, in the Orkneys.

Ayton (Borders) 'Place on the river Eye'; the river-name is *éa* (Old English) 'running stream', with *-tūn* (old English) 'farmstead'; recorded as Eitun, 1098.

B

Back (Western Isles) 'Hollow', from *bac* (Gaelic) 'hollow',
'dip in the ground'.

Badachro (Highland) 'Place of saffron'. *Bad* (Gaelic)
'particular place', *chròch* 'saffron'. *Chro-chorcur* is the
saffron crocus.

Badenoch (Highland) 'Drowned or marshy land'. *Bàithte*
(Old Gaelic) 'liable to flooding', with the suffix *-ach* (Old
Gaelic) 'land'. Noted as Badenach, 1229.

Badentarbet (Highland) 'Place of the isthmus'. *Bad*
(Gaelic) 'particular place' *an* 'of the', *tairbeart* 'isthmus',
'portage place'.

Bal- The most common placename prefix in Scotland, from
Gaelic *baile* 'village', 'home', 'farm' or simply 'place'.
Found throughout the country, except for the extreme
south-east and the Northern Isles. In the east and north,
it may often have supplanted Pictish *Pit-* (see Pit-).

Balaclava (Renfrewshire) Dating from 1856, it is named
after the Crimean War battlefield; there are a number of
other commemorative Balaclavas and also Waterloos as
local names.

Balado (Perth & Kinross) 'Long place'. *Baile* (Gaelic)
'homestead', 'hamlet', and *fhada* 'long'. 'House at the
yew-tree ford', from *baile* with *àth* 'ford' and *eo* 'yew', has
also been suggested.

Balblair (Highland) 'Village on the plain'. *Baile* (Gaelic)
'homestead', 'hamlet', *blàr* 'plain'.

Baldragon (Dundee) Perhaps 'place of the champion'.
Baile (Gaelic) 'homestead', *dreagan* 'dragon', used
figuratively to describe a great warrior.

Balerno (West Lothian) 'Village of the sloe tree'. *Baile* (Gaelic) 'homestead', 'hamlet', *airneach* 'sloe tree'. First recorded in 1280 as Balhernoch.

Balfron (Stirling) Perhaps 'hill house'. Gaelic *baile*, 'homestead', with Brittonic *bronn* 'rounded hillside', 'breast'.

Balintore (Highland) 'Place of the bleaching ground'. *Baile* (Gaelic) 'homestead', 'hamlet', *an* 'of the', *todhair* 'bleaching green'.

Balivanich (Western Isles) 'Monk's place'. *Baile* (Gaelic) 'homestead', 'hamlet', *a'* 'appertaining to', *mhanaich* 'monk's'. The Gaelic name is Baile a' Mhanaich.

Ballachulish (Highland) 'Village of the narrows'. *Baile* (Gaelic) 'homestead', 'hamlet', *caolas* 'narrows', 'strait'. Noted as Ballecheles in 1552.

Ballantrae (South Ayrshire) 'Village on the shore'. *Baile* (Gaelic) 'homestead', 'hamlet', *an* 'on the', *traighe* 'tidal beach'.

Ballater (Aberdeenshire) Possibly 'broom land'. *Bealaidh* (Gaelic) 'broom', *tìr* 'land'. Alternatively 'pass of the water'. *Bealach* (Gaelic) 'mountain pass'; *dobhar* (Brittonic–Gaelic) 'water'. Recorded as Balader 1704, Ballader 1716.

Ballingry (Fife) 'Village of the cave, or den'. *Baile* (Gaelic) 'homestead', *an* 'of', *garaidh* 'cave'. Noted as Ballyngry, c. 1400.

Ballinluig (Perth & Kinross) 'Township by the hollow'. *Baile* (Gaelic) 'homestead', *an luig* 'towards the hollow', from *lag* 'hollow'. The name is found in other localities.

Balloch (West Dunbartonshire, Highland) 'Pass', or 'gap'. Gaelic *bealach*, 'mountain pass', a reference here to the river-gap linking Loch Lomond to the Clyde. 'Balloch' is a frequent element in other placenames. Balloch by Inverness has however been construed as Gaelic *baile an loch*, 'place by the loch'.

Ballygown (Argyll & Bute) 'Place of the smith'. *Baile* (Gaelic) 'homestead', 'hamlet', *ghobhainn* 'smith'. One of the few Bally- names in Scotland, indicating an early name given by the Dàl Riadan migrants around the turn of the 6th century.

Balmaclellan (Dumfries & Galloway) 'MacLellan's place'. *Baile* (Gaelic) 'homestead'; *mac Gille Fhaolain*, (Gaelic patronymic) 'son of the follower of (St) Fillan'. Found in 1183 as Balmacglenin.

Balmaha (Stirling) '(St) Maha's place'. *Baile* (Gaelic) 'homestead', *Mo-Thatha* (Gaelic form of Irish *Tua* 'the silent one', with the prefix *mo-* indicating 'dear'); perhaps a hermit-saint.

Balmerino (Fife) '(St) Merinac's place'. *Baile* (Gaelic) 'homestead'. Recorded as Balmerinach, c. 1200.

Balmoral (Aberdeenshire) Perhaps 'homestead in the big clearing'. *Baile* (Gaelic) 'homestead', *mór* (Gaelic) 'big', *ial* (Brittonic–Pictish) 'clearing'. *Mòrail* (Gaelic) 'majestic', 'splendid' has also been suggested, but the castle's royal ownership dates only from 1852.

Balornock (Glasgow) 'Louernoc's place'. From Brittonic *bod* 'place', and *Louernoc*, a personal name meaning 'little fox'. The earliest form (12th century) is Budlornac.

Balquhidder (Stirling) Perhaps 'fodder store'. *Baile* (Gaelic) 'homestead'; *foidir* (Gaelic form of Old Norse *fothr*) 'fodder'. Its earliest known form is Buffudire (1266), indicating that the prefix was either replaced by, or for a time interchangeable with, *buth* (Gaelic) 'house', 'hut'.

Balta (Shetland) Perhaps 'belt island', from Old Norse *balti* 'belt', and *ey* 'island'.

Banavie (Highland) A stream name, perhaps 'Pig's burn'. *Banbh* (Gaelic) 'pig', *aidh* (Gaelic suffix found with stream names).

Banchory (Aberdeenshire; Perth & Kinross) 'Horns'. *Beannach* (Gaelic) 'horned, forked'. The Gaelic form of the name is Beannachar. The *-y* ending has been explained as the dative form *Beannchraigh*, 'by the bends' of the River Dee. Banchory Devenick and Banchory Ternan relate to churches dedicated to the Celtic saints Devinicus and Ternan. Recorded as Benchorin, 1164; Banchery Defnyk, c. 1300.

Banff (Aberdeenshire) The origin is uncertain. Suggestions include the tentative connection with a traditional Gaelic name for Ireland, *Banba*; and Gaelic *banbh*, 'pig', an animal often met in Celtic mythology. Medieval documents record Banb 1150, Banef 1160, Bamphe 1290, Banffe 1291.

Bangour (West Lothian) 'Goat peak'. *Beinn* (Gaelic) 'mountain', *ghobhar* 'goats'. Shown as Bengouer in the early 14th century.

Bannockburn (Stirling) Bannock- has been related to Brittonic *bannauc*, 'peaked hill', found in the 6th-century 'Gododdin' poem. Scots 'burn' is either an addition, or a replacement of an earlier Brittonic stream-suffix. Recorded as Bannokburne, 1654.

Barassie (South Ayrshire) 'Summit of the droving stance'. *Barr* (Gaelic) 'crest', *fasadh* 'stance', 'level place'. Also found as Barrassie.

Barbaraville (Highland) 'Barbara's town'. One of the numerous late 18th- early 19th-century villages created by lairds and named in compliment to their wives; other examples are Jemimaville and Arabella, in the Black Isle and Easter Ross respectively.

Barlanark (Glasgow) 'Bare hill or ridge'. *Barr* (Brittonic) 'crest', and *lanerc* 'clear space', 'glade'.

Barlinnie (Glasgow) 'Hilltop by the pool'. *Barr* (Gaelic) 'summit', 'top', *linne* 'pool'.

Barnton (East Lothian) 'Barn farm'. Old English *berne* 'barley store', and *tūn*, 'farmstead'. In the Anglian period, barley appears to have been the main cereal crop on the Lothian coast (see Berwick).

Barra (Western Isles) Perhaps 'Isle of St Barr'. Barr or Finnbarr (c. 560–615) from Irish Gaelic *fionn* 'white', and *barr* 'crest'. Alternatively, Gaelic *barr* 'hill', with the later addition of Old Norse *ey* 'island'. The name c. 1200 was Barey; in Gaelic it is Barraidh.

Barrhead (East Renfrewshire) 'Hilltop'. *Barr* (Gaelic) 'crest', 'top'.

Barskimming (East Ayrshire) 'Simon's heights'. *Barr* (Gaelic) 'crest', *Simi* (Gaelic personal name) 'Simon'. Recorded as Barskinning in 1639.

Barry (Angus) 'Barrows' has been suggested, from *beorg* (Old English) with the sense of both 'hill' and 'grave-mound'. But perhaps, like Barry in South Wales, it has a Brittonic source in *barr* 'hill' with *-i*, suffix indicating 'stream'.

Bass Rock (East Lothian) 'The brow'. *An* (Gaelic) 'the', *bathais* 'brow', 'forehead', referring to the distinctive shape of the rock.

Bathgate (West Lothian) 'Boar wood' or 'House in the wood'. *Baedd* (Brittonic) 'boar', *coed* 'wood'. Alternatively, the first element could be derived as *bod* (Brittonic) 'house'. Recorded around 1160 as Bathchet.

Bearsden (East Dunbartonshire) 'wild boars' valley'. *Bár* (Old English) 'boar', *denu* 'valley'. There are no 'bear' names in Scotland.

Beattock (Dumfries & Galloway) Probably 'sharp-topped (hills)'. *Biodach* (Gaelic) 'sharp-topped'.

Beauly (Highland) 'Beautiful Place'. *Beau* (French) 'beautiful', *lieu* 'place'. Recorded in 1230 as *Prioratus de Bello Loco* (Latin) 'priory of the lovely spot'. In Gaelic it is A' Mhanaichann, 'the monastery'.

Beeswing (Dumfries & Galloway) There was an inn here, named after the famous 1830s racehorse, Beeswing.

Beith (North Ayrshire), Renfrewshire) '(place of the) birch tree'. *Beithe* (Gaelic) 'birch tree'. The Renfrewshire name may be Brittonic *beith* 'pig'.

Beley (Fife) A link with the Celtic deity Bel has been proposed, but it appears to be 'farm of the rock-face' from Gaelic *baile* 'homestead', *aill* 'rock', *-in* suffix indicating place, often reduced to -ie. Its oldest form is Balealin, 1220.

Bellahouston (Glasgow) 'Settlement of the crucifix'. *Baile* (Gaelic) 'settlement', *ceusadan* 'crucifix'. Noted as Ballahaustane, 1578. It is assumed that the place possessed a holy cross. In latter centuries the name became confused with Houston.

Ben From Gaelic *Beinn* 'mountain'. The original sense of the word was 'horn', probably first applied to peaked mountains. It has become the standard word for a Scottish mountain, because of its application to the highest peaks. It is also found in the form Bin.

Benarty (Fife) 'Stony hill'. *Beinn* (Gaelic) 'mountain', *artaich* 'stony'. The two parts of the name appear to have been welded togther in common usage.

Benbecula (Western Isles) 'Hill of the salt-water fords'. The Gaelic name is Beinn na Faoghla. *Beinn* (Gaelic) 'mountain', *na* 'of the', *faoghail* (adapted Gaelic form of *fadhail*) 'salt-water ford'. Alternatively, the last element may be *faoghlach* (Gaelic) 'strand', 'beach'. Recorded as Beanbeacla 1495; Benvalgha 1549; Benbicula 1660.

Benderloch (Argyll & Bute) 'Hill between two lochs'. Gaelic *Beinn* 'mountain', *eadar* 'between', *dá* 'two', *loch* 'lake', 'loch'.

Bennachie (Aberdeenshire) 'Mountain of the nipple'. *Beinn* (Gaelic) 'mountain; *na* 'of', *cìche* 'nipple, breast'. The derivation 'blessed place', from *beannaichte* (Gaelic) 'blessed', has also been suggested, but seems less likely.

Berneray (Western Isles) 'Bjorn's isle'. *Bjorn* (Old Norse personal name), *ey* 'island'. The Gaelic name is Eilean Bhearnaraigh. Bernera is the same.

Berwick upon Tweed (Northumberland) 'Barley farm.' *Bere* (Old English) 'barley', 'bere', *wic* 'farm'. The name is found as Berwic from 1095. See Tweed.

Bettyhill (Highland) Named after Elizabeth, countess of Sutherland, this village was set up about 1820. In Gaelic it retained its name of Am Blàran Odhar, 'the brown fields'.

Biggar (South Lanarkshire) 'Barley field'. *Bygg* (Old Norse) 'barley', *gardr* 'enclosure'. Alternatively, the latter may be the more specific *geiri* (Old Norse) 'triangular plot'. Early records include Bigir, 1170, Begart, 1524.

Birnam (Perth & Kinross) 'Rainy place' from Gaelic *braon* 'rain', 'shower', with -*an* locative suffix. Noted as Brannan, 1345. 'Birnam wood' in *Macbeth* perpetuates a mistaken form.

Birsay (Orkney) 'Hunting-ground valley'. *Birgis* (Old Norse) 'hunt', *herath* 'valley'. The name is found as Birgisherad in the *Orkneyinga Saga*, c1225.

Bishopbriggs (East Dunbartonshire) 'Bishop's lands'. The 'bishop' is that of Glasgow; *riggs* (Scots) 'fields'. The *b* has crept in through confusion with Scots *brig* 'bridge'.

Black Isle, The (Highland) Perhaps an English translation of Gaelic *eilean* 'island', *dubh* 'dark', referrring either to black soil or dark forest cover. An alternative explanation is that the Gaelic name was Eilean Duthac, St Duthac's island, mis-translated into English as 'black isle'.

Blackwater, River (Highland; Perth & Kinross; Stirling) Translation of Gaelic *allt* 'stream' and *dubh* 'dark'. This is a frequently found river-name.

Blair Atholl (Perth & Kinross) 'Plain of the New Ireland', or 'plain of the ford of Fodla'. *Blàr* (Gaelic) 'plain', 'level clearing', often with the sense of 'battlefield'; and see Atholl.

Blairgowrie (Perth & Kinross) 'Plain of Gabran'. *Blàr* (Scottish Gaelic) 'plain', 'level clearing'; the second element relates to the district of Gowrie, perhaps to distinguish it from Blair Atholl. Noted as Blair in Gowrie, 1604. *See* Gowrie, Rattray.

Blantyre (South Lanarkshire) 'Edge-land'. *Blaen* (Gaelic, from Brittonic) 'edge'; *tìr* (Gaelic) 'land'. Noted in 1289 as Blantir.

Boat of Garten (Highland) 'Boat' as an inland name indicates the one-time existence of a ferry, in this case across the River Spey. The Gaelic name is Coit Ghartainn, from *coit*, 'small boat'. See Garten.

Boddam (Aberdeenshire; Stirling) 'Bottom place'. Old English *botm* 'bottom'. Boddam is placed at the cliff-foot. Buddon, as in Buddon Ness, and Boddin are cognate.

Bogie, River and **Strath** (Aberdeenshire) Perhaps 'stream with the bag-like pools', from *balg* (Gaelic) 'bag', and *aidh* stream-related suffix, perhaps a Pictish borrowing. Earlier forms include Strabolgin, 1187.

Bonar Bridge (Highland) 'Bridge at the lowest ford'. *Bonn* (Gaelic) 'bottom', 'base'; *àth* 'ford'.

Bonawe (Argyll & Bute) 'Water-foot'. *Bonn* (Gaelic) 'bottom', and *abh* (Old Gaelic) 'water'. The Gaelic name is sometimes given as *Bonn-atha* 'ford at the foot' (*see* Bonar).

Bo'ness (West Lothian) A contraction of Borrowstounness, Old English *Beornweard* (personal name) with *tūn*, 'farmstead': 'Beornweard's farm', later assimilated to Borrowstoun (Scots) 'burgh', 'town'; *naes* (Old English) 'promontory'. Noted as Berwardeston, c. 1335.

Bonhill (West Dunbartonshire) 'Hut on the hill'. Old English *bothle*, 'hut', and *hyll* 'hill'. Found c. 1270 as Buthelulle.

Bonnybridge (Falkirk) 'Swift stream bridge'. *Buan* (Gaelic) 'swift'. The name originally referred to the stream, now reduced by back-formation to being the 'Bonnybridge Burn'.

Boreraig (Highland) 'Fort bay'. *Borgar* (Old Norse) 'fortified site'; *aig* (Gaelic from Old Norse *vik*) 'bay'.

Borgue (Western Isles, Highland, Dumfries & Galloway) 'Fort'. *Borgar* (Old Norse) 'fortified site'.

Borrodale (Highland) 'Dale of the fort'. *Borgar* (Old Norse) 'fortified site', *dalr* 'valley'. A superfluous Glen- has been added; the -*dalr* ending never having been borrowed into Gaelic.

Borve (Western Isles) 'Fort'. *Borgar* (Old Norse) 'fort'. This name, found in several of the Outer Hebrides, is related to Burgh and Brough in Shetland and Orkney.

Bothwell (South Lanarkshire) This been taken as 'Buth's pool'. *Buth* (Old English personal name) *waelle* 'pool'; the first part has also been seen as Old English *bothe* 'hut', 'house'. Noted in 1242 as Botheuill.

Bower (Highland) 'House', from Old Norse *búr*, 'house'. Noted as Bouer, c. 1230.

Bowling (West Dunbartonshire) Perhaps 'the place of Bolla's people.' *Bolla* (Old English personal name), *inga* 'of the people of'. A Gaelic source has also been suggested, in *Bò* 'cow's', *linn* 'pool'. In 1303 it was recorded as Bolline.

Bowmore (Argyll & Bute) Place of the 'big house'. *Both* (Gaelic) 'house', *mór* 'big'.

Braan, River and **Strath** (Perth & Kinross) 'Roaring stream'. *Breamhainn* (Gaelic) from an older Celtic root **bremava* 'noisy', 'rumbling'. Noted in 1200 as Strathbranen.

Braco (Stirling) 'Grey place'. *Braca* (Gaelic) 'grey', 'greyish', *-ach* affix denoting place.

Braemar (Aberdeenshire) 'The upper part of Mar'. *Bràigh* (Gaelic) 'upland'; the second element, *Mar*, is a personal name of unknown derivation. Early records include Bray of Marre 1560, Brae of Mar 1610, Breamarr 1682. In the 19th century it was known as Castleton of Braemar.

Braemore (Highland) 'Big upland'. *Bràigh* (Gaelic) 'upland', *mòr* 'big'.

Braeriach (Moray) 'Brindled or grey hill'. Gaelic *bràigh* 'upland', borrowed into Scots as 'brae', *riabhach* 'brindled', 'greyish'.

Braes (Highland and other regions) 'Uplands'. *Bràigh* (Gaelic) 'upland'.

Brander, Pass of (Argyll & Bute) 'Brander' may mean 'pot', from Gaelic *brannraidh* 'pot', though older Gaelic meanings also include 'trap' or 'snare'.

Braid Hills (Midlothian) 'Upper slopes', or 'gullet'. *Bràghad* (Gaelic) 'neck'.

Breadalbane (Perth & Kinross; Stirling) 'Upper part of Alban'. *Bràghad* (Gaelic) 'higher, upper part', 'hill district', *Albainn* 'of Scotland'. The *Pictish Chronicle*, c. 970, refers to Brunalban, from Brittonic–Pictish *bryn*, mountain.

Breacleit (Western Isles) 'Broad cliff or hill'. *Breidhr* (Old Norse) 'broad', *klettr* 'rocky holm', 'cliff'. Breascleit, also in Lewis, is from Old Norse *breid-áss-klettr*, 'broad ridge cliff'.

Breich (West Lothian) A stream name, possibly Gaelic *breac* 'trout' or Brittonic *brech* 'speckled.

Brechin (Angus) 'Brychan's place'. Perhaps after the Celtic hero Brychan. Old forms are found in the genitive, as in the *Pictish Chronicle*, c. 970, *civitas brechni*, 'Brychan's community'. But a Pictish ruler, Brachan, has been noted in the Angus region: it may be 'Brachan's community'.

Bressay (Shetland) Probably 'breast-shaped island'. *Brjost* (Old Norse) 'breast', *ey* 'island'.

Brig o' Balgownie (Aberdeenshire) 'Bridge of the smith's place'. *Brig* (Scots) 'bridge'; *baile* (Gaelic) 'place', 'homestead', *gobhainn* 'of the smith'. The bridge was built in 1329, when Scots was the language of the area.

Brig o' Turk (Stirling) 'Bridge of the pig'. *Brig* is the Scots translation of *droichead* (Gaelic) 'bridge', *nan* 'of', *tuirc* 'hog', boar'.

Brittle, Glen (Highland) 'Broad bay'. *Breithr* (Old Norse) 'broad', *vik* 'bay'. The 'Glen' is a later addition to designate the valley behind the bay.

Broadford (Highland) 'The broad ford', English translation of Gaelic *an t-àth* 'the ford', *leathan* 'broad'. *An t-àth Leathann* remains the Gaelic name.

Brodick (North Ayrshire) 'Broad Bay'. *Breithr* (Old Norse) 'broad', *vik* 'bay'. Early records show: Brathwik 1306; Bradewik 1488. In Gaelic it is Breadhaig.

Brodie (Moray) 'Place by the ditch' or 'muddy place'. *Brothag* (Gaelic) 'ditch', 'hollow'.

Brogar (Orkney) Perhaps 'Place by the bridge'; *Bru* (Old Norse) 'bridge', *gardr* 'enclosure', 'garth', but the locality name is Brodgar, which may stem from Old Norse *breith* 'broad', and seems more apt to the site.

Broom, River, Loch (Highland) 'Falling water'. *Braon* (Gaelic) 'water', 'drop'. The upper loch, retaining its Gaelic name, Loch a' Bhraoin, and the sea-loch take their name from the river linking them. Loch Braoin in Perthshire has the same derivation.

Broomielaw (Glasgow) 'Hill of broom'. *Law* (Scots) 'hill.'

Brora (Highland) 'Place of the bridge's river'. *Bru'r* (Old Norse) 'bridge', *aa* 'river'. Noted in 1499 as Strabroray, 'Strathbrora'. Unusually, a river has been named after a bridge here, presumably because of the general rarity of bridges.

Broughton (Borders; Midlothian) 'Brook place'. *Brōc* (Old English) 'brook', *tūn* 'village', 'farmstead'. A rare use of *brōc*: 'burn' from O.E. *burna*, predominates. Noted as Broctuna, 1128.

Broughty Ferry (Dundee) 'Tay-bank ferry'. *Bruach-taibh* (Gaelic) 'bank of the Tay', provides the first element. Recorded as Bruchty Craig, 1541.

Broxburn (West Lothian) 'Badger's stream'. *Brocc-s* (Old English) 'badger's', *burna* 'stream'.

Buachaille Etive (Highland) 'The herdsman of Etive'. *Buachaille* (Gaelic) 'herdsman', *Eite* 'of Etive'. An unusually imaginative Gaelic hill-name, but they have older names as Stob Dearg ('red peak') and Stob Dubh ('black peak'). There are two 'herdsmen', *Mór* 'big', and *Beag* 'small'. See also Etive.

Buchan (Aberdeenshire) Possibly 'place of the cow'. *Buwch* (Brittonic, with the suffix -*an*) 'cow place'. An alternative derivation is from Gaelic *bogha* 'bend', 'angle'. Recorded as Buchan around 1150 in the *Book of Deer*, Baugham 1601.

Buchanan (Stirling) May be related to Buchan, or perhaps 'priest's house' from *both* (Gaelic) 'house', *chanain* 'priest'. Noted as Buchquhanane around 1240.

Buchlyvie (Stirling) 'House or hut on the slope'. *Both* (Gaelic) 'hut', *slèibhe* 'slope'.

Buckhaven (Fife) Dating from the mid-16th century, probably of similar derivation to Buckie, 'harbour of bucks or buckies'.

Buckie (Moray) Probably 'Place of bucks'. *Bocaidh* (Gaelic) 'whelk'. *Bucaidh* (Gaelic) 'pimple', 'protuberance', has also been suggested. Bucksburn, Aberdeen, would seem to suggest the latter meaning. Recorded as Buky c. 1350.

Bunessan (Argyll & Bute) 'Foot of the waterfall'. *Bonn* (Gaelic) 'bottom', *easan* 'little waterfall'.

-burgh This suffix can stem from Old English *burh* or Old Norse *borgr*, both originally meaning 'fortified settlement'. In the O.E. cases it has developed as Scots 'burgh', a town with a charter, possessing certain rights. The O.N. form has largely retained its original sense.

Burghead (Moray) 'Headland of the fort'. *Borgr* (Old Norse) 'fort'. 'Head' here may be a Scots translation of *nes* (Old Norse) 'headland'.

Burntisland (Fife) Experts now consider the name to be literal, following the 16th century burning of buildings to incorporate a small island into the new harbour. In 1538 it was noted as Scots 'Brint Eland'.

Burra (Shetland) 'Fortified isle'. *Borgar* (Old Norse) 'fort', *ey* 'island'. Burray (Orkney) has the same derivation.

Busby (East Renfrewshire) 'Bushy place'. *Busk* (Old Norse) 'bush', and *by* 'farm' or 'place'. Noted c. 1300 as Busbie.

Bute (Argyll & Bute) Possibly 'area of land'. *Bót* (Old Norse) 'patch' or 'piece of land'. A derivation from *bot* (Irish Gaelic) 'fire', has also been suggested. Records show Bot 1093, Bote 1204, Boot 1292. The modern Gaelic name is *Bód*.

Butt of Lewis (Western Isles) The only cape so named, and its origins are unclear. Old French *buter*, 'to butt out', has been suggested, as has Danish *but*, 'stumpy', as a root. There may be some connection with the Buddons/Buttons of eastern promontories. Its Gaelic name is Rudha Robhanais, *rudha* 'point'; *raud* (Old Norse) 'red', *nes* (Old Norse) 'point'.

C

Cadboll (Highland) Perhaps 'place of wild cats', though the cats may be tribal symbols. Noted as Kattepoll, 1281; Cathabul, 1529. *Cat* (Gaelic) 'cat'; *ból* (Old Norse abbreviated form of *bolstadr*) 'farm'.

Cadzow (South Lanarkshire) This previous name for Hamilton may commemorate a battle, from *cath* (Gaelic) 'battle'. The latter part may be cognate with *howe* (Scots) 'hollow'. The name is recorded in 1150 as Cadihou. As in numerous other placenames, the *z* is a relic of the attempt to convey the sound 'gh' in writing, and does not intend alphabetical 'z'.

Caerlaverock (Dumfries & Galloway) Perhaps 'Fort in the elm trees'. *Cathair* (Gaelic) 'fort', *leamh-reaich* 'elm tree'; or Brittonic *caer* 'fort' with a personal name *Lifarch*. The present form is influenced by the phonic resemblance to Scots *laverock* 'skylark'.

Cairndow (Argyll & Bute) 'Dark mount'. *Càrn* (Gaelic) 'humped hill', *dubh* 'dark'.

Cairn Gorm (Moray; Highland) 'Blue humped hill'. *Càrn* (Gaelic) 'humped hill', *gorm* 'blue', 'green'. This single mountain's name, run into one as 'The Cairngorms', has been adopted for the whole range. The Gaelic name is *Monadh Ruadh* 'red mountains'.

Cairnie (Moray) 'Place of thickets'. *Cardden* (Pictish) 'thicket'; *ach* (Gaelic) locative suffix. Also Cairney. It is frequently found in other areas as a farm or field-name.

Cairnryan (Dumfries & Galloway) 'Fort by (Loch) Ryan'. *Caer* (Brittonic) 'fort', and see Ryan.

Caithness (Highland) The first element is noted in the *Pictish Chronicle*, c. 970, as a Pictish province-name, *Cat* or *Cait*; supposedly from a son of Cruithne, 'founder' of the Picts. The suffix, added in the 9th-10th century by Norse colonists, is *nes* (Old Norse) 'headland'. The *Book of Deer* notes Catness, c. 1150, suggesting that the Gaelic name had not yet become *Gallaibh*, 'land of the strangers'.

Calder (Highland; West Lothian; Argyll & Bute) Probably 'hard or rapid-flowing water'. *Cal-* indicates 'hard' in the Celtic languages, with *dobhar* 'water'. Old Gaelic *call*, 'hazel' has also been suggested for the first part. Noted in Lothian as Kaldor, c. 1250, Caldovere, 1293. Calder in Caithness has been linked to Old Norse *kálfr* 'calf', *dalr* 'dale', found as Kalfadal in the *Orkneyinga Saga*, c. 1225.

Caldercruix (North Lanarkshire) 'Bends of Calder'. *Cruix* (Scots) from Old Norse *krókr* 'crooks', 'bends'.

Caledonia 'Scotland'. Used by the Romans from a tribal name, it may not have a Celtic linguistic origin though it has been linked to Brittonic *caled*, 'with the quality of hardness'. The tribe of Caledones is first noted on Ptolemy's map (CE150).

Calgary (Argyll & Bute) 'Cali's garth'. *Kali* (Old Norse personal name), *gerdhi* (diminutive of *gardr*) 'garth': the land between the machair and the moorland, taken into Gaelic as *gearraid*.

Callander (Stirling) Perhaps 'ferry shore' from Gaelic *caladh* 'ferry', *sraid* 'street', 'firm shore'. The Gaelic name is Calasraid. Other explanations link it to Cal- river names (see Calder). A 1457 source gives *terras de* 'lands of' *Calyn et Calyndrade* where -drade may be Gaelic *tràghad* 'of the shore'. Calyn remains unexplained.

Callanish (Western Isles) 'Kali's ness'. *Kali* (Old Norse personal name), *nes* 'cape', 'point'. Also found as Callernish.

Calton (Edinburgh; Glasgow and other areas) 'Hazel (place)'. *Calltuinn* (Gaelic) 'hazel'. The numerous local Caldon placenames in the south-west have the same source. Some post-Gaelic Caltons might be 'cold place', from Scots *cauld*, 'cold', and *toun*, 'farmstead'.

Cambus (Stirling) 'Place of the bay'. *Camus* (Gaelic) 'bay', 'beaching place'. Most *camus* names have some further identifier, but not here.

Cambuskenneth (Stirling) 'Cinaed's Bay'. *Camus* (Gaelic) 'bay'; the second part may be a Scots form of the Pictish personal name *Cinaed*, or of Gaelic *Choinnich* Kenneth. Found in this form 1147; Cambuschynoch 1527.

Cambuslang (South Lanarkshire) 'Bay of the ship'. *Camus* (Gaelic) 'bay', *luinge* 'ship's'. Found as Cameslank, 1296.

Cambus o' May (Aberdeenshire) 'Bay on the flat land'. *Camus* (Gaelic) 'bay'. May has been inferred to come from Old Irish *Miathi*, a form of Maeatae, a tribe identified by the Romans; but more probable is *magh* (Gaelic) 'plain', on a bend of the River Dee.

Camelon (Falkirk) Curvy stream. *Cam* (Brittonic) 'bent', 'crooked', *linn* 'stream'. The resemblance to Arthurian Camelot or Camlann is accidental.

Cameron (Fife; Edinburgh) Suggested as 'crooked hill', from Gaelic *cam* 'crooked', and Old Gaelic *brun* 'hill', cognate with Brittonic *bryn*. Found in 1190 as Cambrun.

Campbeltown (Argyll & Bute) Named after Archibald Campbell, earl of Argyll, in 1667. Formerly it had been called Lochhead, and even earlier Kilkerran.

Camperdown (Dundee) Commemorating the naval battle of Camperdown, 1797, won by Admiral Lord Duncan, who was born in Dundee.

Campsie Fells (East Dunbartonshire) 'Crooked hills'. *Cam* (Gaelic) 'crooked', *sith* 'hill'. As *sith* also means 'fairy', the name has sometimes been taken as 'fairy hills'. 'Fells' from *fjall* (Old Norse) 'hill', is a later addition. Recorded as Kamsi 1208; Camsy 1300; Campsy 1522. The *p* is epenthetic (interpolated through pronunciation).

Canisbay (Highland) 'Canons' place. Shown around 1240 as Cananesbi, with *canane* (early Scots) 'canon'; and *by* (Old Norse) 'village'.

Canna (Highland) 'Bucket island'. *Kanna* (Old Norse) 'can', 'bucket'; presumably from the island's shape.

Canonbie (Dumfries and Galloway) 'Canons' village'. *Canon* (Middle English) refers to the Augustinian priory that was here, 1165–1542; -*by* (Old Norse) 'village'. Noted as Canneby, 1290.

Cannich, River, Glen (Highland) Perhaps 'place where cotton sedge grows'. *Canach* (Gaelic) 'cotton sedge'. The Gaelic name is Canaich.

Caputh (Perth & Kinross) 'Hill top'. *Ceap* (Gaelic) 'hilltop', with -*ag* diminutive ending. Related to Latin *caput*, 'head'.

Carbost (Western Isles; Highland) 'Kari's steading'. *Kari* (Old Norse personal name), *bolstadr* 'farm'.

Cardenden (Fife) 'Den or hollow of the thicket'. Pictish *cardden* 'thicket'; *den* (Scots, from Old English *denu*) 'small steep valley'.

Cardrona (Borders) Possibly 'fort of winds or breezes'. *Cathair* (Gaelic) 'fort', *drothanach* 'breezy'. Noted as Cardronow in 1530.

Cardross (Argyll & Bute, Stirling) 'Wooded promontory'. *Cardden* (Pictish) 'thicket'; *ros* may be Gaelic 'promontory' or Brittonic 'moor'. Recorded in 1208 as Cardinros.

Carlops (Borders) 'The hag's leap'. *Carline* (Scots) 'old woman'; *loups* (Scots) 'leaps'. In Wyntoun's *Chronicle* (c. 1400) it is referred to as Karlinlippis. The name was first given to the stream location, the village being founded in 1784.

Carloway (Western Isles) 'Karl's Bay'. *Karla* (Old Norse personal name), *vágr* 'bay', gaelicised into *Càrlabagh*.

Carluke (South Lanarkshire) Probably 'Fort on the marsh'. *Caer* (Brittonic) 'fort', *lwch* 'marshland'. First recorded in 1320 as Carneluke.

Carmyllie (Angus) 'Warrior's fort'. *Caer* (Brittonic) 'fort'; *milidh* (Gaelic) 'warrior'.

Càrn Smeart (Highland) A Ross-shire hill that appears to preserve the name of the Smertae tribe, recorded in this area by Ptolemy, c. CE 150; *càrn* (Gaelic) 'humped hill', *Smeart* related to Gaelic *smior* 'marrow', 'courage'.

Carnoustie (Angus) Possibly 'rock of the fir tree'. *Carraig* (Gaelic) 'rock', *na* 'of the', *ghiuthais* 'fir tree'. A late 15th century document records Carnusy. The *t* was interpolated at a later date. If the first element is Gaelic *càrn* 'hill', the second may be like Ussie in Ross-shire, described as "obscure, Pictish or pre-Pictish".

Carnwath (South Lanarkshire) Probably 'New fort', from Brittonic *caer* 'fortified site' and *nowith* 'new'. Recorded around 1179 as Karnewid.

Carrbridge (Highland) 'Bridge at the rocky shelf'. *Drochaid* (Gaelic) translated to English 'bridge', *charra* 'at the rock shelf'.

Carrick (South Ayrshire) 'Rocky place'. *Carraig* (Gaelic) 'rock', 'crag'.

Carron (Falkirk; Highland; Moray; Dumfries & Galloway) 'Rough river.' The numerous Carron names derive from the early-Celtic root **kar* 'harsh', 'rough', that gives Gaelic *càrn*, 'heap of stones'. The *-on* suffix is found in other ancient river-names, from a root-form **ona*, indicating 'water'. The Falkirk Carron is found as Caere in 710, Carun in the 9th century.

Carsphairn (Dumfries & Galloway) 'Carseland of the alder-trees'. *Carse* (Scots from Old Norse *kerss*) 'low-lying land by a river;' *feàrna* (Gaelic) 'alders'. Above the village rises Cairnsmore (from Gaelic *càrn* 'hill', and *mòr* 'big') of Carsphairn.

Carstairs (South Lanarkshire) 'Castle Tarras'. *Caisteil* (Gaelic) 'castle'; *Tarras* is probably a personal name. Recorded as Casteltarres in 1170.

Cart, River (Renfrewshire) Perhaps 'cleanser'. The resemblance of the name to Gaelic *càraid*, 'pair', has been noted since this river is formed by the joining of the White and Black Cart streams. A link with an older root-form, Old Irish Gaelic *cartaim* 'I cleanse', or with the same pre-Celtic root-form, **kar* 'hard', 'stony', as Carron, seems more probable.

Cassillis (North Ayrshire) 'Castles', from Irish Gaelic *caiseal* 'stone fort'. Found as Casselys, 1385. A number of similar names are also linked to sites of brochs or forts, including Cashel Point on Loch Lomond, Cashlie in Glen Lyon, and Glen Cassley in Ross-shire.

Castlebay (Western Isles) The name refers to the MacNeil castle of Kisimul. The Gaelic form is *Bagh a Chaisteil*, though the usual name was *Baile MhicNéill*, 'MacNeil's place'.

Castle Douglas (Dumfries & Galloway) 'Castle of the Douglas family'. This town was developed by Sir William Douglas in 1789. Previously it was known as Causewayend, then Carlingwark (perhaps Scots 'work of the hag').

Castlecary (North Lanarkshire) 'Fort of the fortifications'. 'Castle' is late 12th century, added when the sense of *caerydd* (Brittonic) 'forts', was lost: Castlecarris, c. 1200.

Castlemilk (Dumfries & Galloway, Glasgow) 'Castle on the milky stream'. 'Milk' may be from Old English *meolc*, a stream name with colour associations. The Milk Water is a tributary of the Annan, the name transferred when the proprietors acquired land near Glasgow, c1470.

Cathcart (Glasgow) 'Wood' or perhaps 'fort' of Cart. *Coet* (Brittonic) 'wood', or *caer* 'fort', with *Cert*, a river-name, see Cart. Old forms include Kerkert, 1158, and Katkert, c. 1170.

Cathkin (South Lanarkshire) 'Place of common pasture'. *Coitchionn* (Gaelic) 'common pasture'.

Catrine (East Ayrshire) 'Battle point' has been suggested. *Cath* (Gaelic) 'battle', *roinne* 'point'.

Catterline (Aberdeenshire) 'Fort by the pool'. *Cathair* (Gaelic) 'fort', *linne* 'pool'. Noted in 1201 as Katerlen.

Cawdor (Highland) 'Hard or fast-flowing water'. The derivation is probably the same as for Calder, with a form Kaledor noted in 1280. Its Gaelic form is Caladair.

Ceres (Fife) 'Western place'. *Siar* (Gaelic) 'western', *-ais* 'place'. Found as Syreis in 1199.

Cessford (Borders) 'Cessa's enclosure'. *Cessa* (Old English personal name); *worth* (Old English) 'enclosed fields', altered into 'ford'. Anglo-Saxon *-worth* names, common in England, are rare in Scotland; by the time Anglians were moving in, it was obsolete, and *-tūn* or *-ham* were the normal forms.

Challoch (Dumfries & Galloway) 'Anvil'. *Teallach* (Gaelic) 'anvil', 'forge'.

Chesters (Borders) 'Camps', from Latin *castrum*, 'military camp', giving Early English *ceaster*. There are a number of other Roman-related names in this region: Bonchester Bridge appears to combine *bonn* (Gaelic) 'foot', with *ceaster*.

Cheviot (Borders) The source is unclear; it may be linked to a pre-Celtic tribal name. An early form of Brittonic *cefn* 'ridge' is likely. It is noted as Chiuiet, 1181.

Clachan (Highland; Argyll & Bute) 'Place of stones'. *Clachan* (Gaelic) 'stones'.

Clachtoll (Highland) 'Hollow of stones'. *Clach* (Gaelic) 'stone', *toll* 'hollow', 'hole'.

Clackmannan (Clackmannanshire) 'Stone of Manau or Manan'. *Clach* (Gaelic) 'stone' perhaps replacing a Brittonic term relating to Welsh *clog* 'rock'. *Manau* or *Manan* is an ancient personal or divinity name given to this area. Found in present form in 1221.

Clashmore (Highland) 'Deep furrow'. *Clais* (Gaelic) 'trench', 'furrow', *mòr* 'big'.

Cleghorn (South Lanarkshire) 'Clay house'. *Claeg* (Old English) 'clay', *erne* 'house'. Noted in 1230 as Clegerne.

Cleish (Perth & Kinross) 'Furrow', 'trench', from Gaelic *Clais*. The form Kles is recorded in 1221.

Clett 'Cliffs', from Old Norse *klettr*. There are many Clett names in Shetland and Orkney. Clatt in Aberdeenshire is also derived from *clett*. *See also* Breacleit.

Clickhimin (Shetland; other districts) 'Rock mouth', from Old Norse *klakk* 'rock', and *minni* 'mouth'. The Cleekhimin Burn flows into the Leader Water in the Borders.

Clisham (Western Isles) 'Craggy hill', from Old Norse *klif-s-hamra* 'crag', and *holmr* 'hill'.

Cloch (Inverclyde) 'By the stone', from Gaelic *cloiche*; noted in 1600 as Clochstane.

Cluanie (Highland) 'Meadow'. *Cluain* (Gaelic) 'meadow', with *-ach* locative ending. Cluny and Clunie have the same derivation.

Clyde, River, Strath, Firth (South Lanarkshire; Renfrewshire; Inverclyde) Probably 'Cleansing one'. A Brittonic river-name, traced back to a hypothetical **clouta* stemming from the Indo-European root element **clut*, with the sense of 'washing'; and so linked in significance with such other river-names as Nith, Nethy, which also convey a sense of purification. Recorded in the first century AD by Tacitus as Clota. Adamnan's *Life of St Columba* refers to *Petra Cloithe* 'rock on the Clyde'. Strathclyde is recorded as Straecled, 875.

Clydebank (West Dunbartonshire) A modern name, given when the town was developed in the 19th century.

Clynder (Argyll & Bute) 'Red slope'. *Claon* (Gaelic) 'slope', *dearg* 'red'. Clinterty in Perthshire, and also near Dyce in Aberdeenshire may have the same derivation, with *taigh* 'house', added.

Coalburn (South Lanarkshire) A modern name from the mid 19th century, reflecting local mining activity.

Coatbridge (North Lanarkshire) 'Bridge by the cottages'. The bridge was built only about 1800 as part of the development of the area around of Cotts (Old English *cot* 'hut,' 'cottage'). It is found as Coitts in 1582.

Coatdyke (North Lanarkshire) 'Cottage by the dyke'. *Cot* (Old English) 'hut,' 'cottage'; *dyke* (Scots) 'wall'.

Cockburnspath (Borders) 'Colbrand's path', from Old English *Colbrand* (personal name), and path. Earlier forms include Colbrandespade (c. 1128); by 1508 Cokburnspath is found but the 'ck' has never been pronounced.

Cockenzie (East Lothian) Derivation not fixed; perhaps 'point of the bonnet' from Gaelic *coc* 'cap' and *eang* 'point of land', 'nook'. Recorded as Cowkany in 1590, and still pronounced as 'Cockennie'.

Coigach (Highland) 'The place of fifths'; *na cóigich* (Gaelic) 'of the fifths': implying a form of land division in times past.

Coldstream (Borders) Before the bridge was built (1766), the Tweed was forded here, and the name may refer to the temperature of the river. Recorded 1128 as Kaldestrem. But *kald* may go back much further, with 'stream' replacing an earlier river suffix. See Calder.

Coldingham (Borders) 'Village of the people of Colud'. *Colud* (Brittonic personal name); *inga* (Old English) affix with the sense of 'people of'; *ham* (Old English) 'village'. Noted around 798 as Coludesburg, with the present form found in 1098.

Colinsburgh (Fife) Named after Colin Lindsay, earl of Balcarres, who founded it in 1705.

Colinton (Edinburgh) 'Colban's farm'. The form in 1319 was Colbanestoun. Colban (Gaelic personal name) may have supplanted an earlier Anglian name: the suffix is from Old English *tūn* 'farmstead'.

Colintraive (Argyll & Bute) 'Swimming narrows', from Gaelic *caol* 'strait', 'kyle', and *snàimh* genitive form of *snàmh* 'swimming' (probably of cattle).

Collessie (Fife) 'Nook of the water', from Gaelic *cuil* 'nook', 'corner', and *eas* 'waterfall', with the locative suffic -*ach*.

Coll (Argyll & Bute) Probably 'barren place'. *Kollr* (Old Norse) 'bald head', 'bare top'.

Colonsay (Argyll & Bute) Possibly '(St) Columba's Isle'. Alternatively, *Kolbein* (Old Norse) as a personal name; the second part is from *ey* (Old Norse) 'island'. Older forms include Golwonche, 1335, Colowsay, 1376.

Coltness (North Lanarkshire) 'Wood by the water (fall)'. *Coille* (Gaelic) 'wood', *an* 'by the', *eas* 'waterfall'.

Comiston (Edinburgh) 'Colman's farm'. *Colmàn* (Gaelic proper name, deriving from *Colum*, 'dove') and *tūn* (Old English) 'farmstead'. Found as Colmanston, 1335.

Comrie (Perth & Kinross) 'Place of the confluence'. *Comar* (Gaelic) 'river confluence'. Noted as Comry, 1268.

Condorrat (North Lanarkshire) 'Place of the confluence'. *Comh-* (Gaelic prefix which can indicate 'convergence'), *dobhar* 'water', 'stream', *-ait* suffix indicating place). Recorded in 1553 as Cundurat.

Connel (Argyll & Bute) 'Whirlpool'. From Gaelic *coingheall*, 'whirlpool', referring to the tidal rapids here at the entrance to Loch Etive

Conon River, Strath (Highland). 'Wolf, or dog, river'. The root form appears to be *con* (genitive form of Old Irish *cu*) 'wolf', 'dog'. The *-ona* (early Celtic) 'water' suffix is frequently found in river-names. Often found in pre-20th century sources as Conan.

Contin (Highland) The root form appears to be, as with Conon, *con* (genitive form of Old Irish *cu*) 'wolf', 'dog', with *dainn* (Gaelic) 'rampart'. Perhaps 'fort of the dog'. Noted as Conten in 1226.

Copinsay (Orkney) 'Kolbein's isle'. From Old Norse *Kolbein* (personal name) and *-ey* 'island'.

Corgarff (Aberdeenshire) 'Rough corrie'. *Coire* (Gaelic) 'corrie', 'mountain hollow', *garbh* 'rough'.

Corpach (Highland) 'Corpse-place'. *Corpach* (Gaelic) 'corpses'. Funerals from Fort William to the church at Annat rested here.

Corstorphine (Edinburgh) Possibly 'cross of the fair hill'. *Crois* (Gaelic) 'cross', *torr* 'hill', *fionn* 'fair'. Less likely derivations include one from Torphin (personal name) or from the eleventh-century Norse earl Thorfinn the Mighty. Records show Crostorfin 1128; Corstorphyne 1508.

Coruisk, Loch (Highland) 'Water hollow'. *Coire* (Gaelic) 'mountain hollow', *uisge* 'water'.

Coshieville (Perth & Kinross) 'By the trees'. *Cois* (Gaelic) 'beside', *a* 'the', *bile* 'thicket'. The modern form has developed by modelling on -ville suffixes.

Coulter (South Lanarkshire) 'The back land'. *Cul* (Gaelic) 'back', *tìr* 'land'. Noted as Cultyr, c. 1210. Also found as Culter.

Coupar Angus (Perth & Kinross) Named to distinguish it from Cupar, Fife, but no longer located in Angus. Perhaps derived as proposed for Cupar, from a lost Pictish word denoting 'confluence'. The Couper Burn joins the Isla here. Found as Cubert, 1169; Coupre in Anegos, 1296.

Cove (Argyll & Bute; Aberdeenshire; Borders; Highland) 'Hut', from *kofi* (Old Norse) 'hut'. Cove Bay in Cowal and south of Aberdeen are 'hut bays'.

Cowal (Argyll & Bute) A corruption of *Comhgall*, the name of a grandson of Fergus of Dàl Riada.

Cowcaddens (Glasgow) 'Hazel nook'. *Cuil* (Gaelic) 'corner', 'nook', *calldainn* 'of hazels'.

Cowdenbeath (Fife) Possibly 'Cowden's (land) by the birches'. *Cowden* (personal name) suggested by the earliest known record in 1626, which refers to the place as Terris de Cowdounesbaithe; the second element is derived from *beith* (Gaelic) 'birch'.

Coylum (Highland) 'Gorge leap'. *Cuing* (Gaelic) 'gorge', *leum* 'leap'. The name encourages speculation as to how wayfarers got across the River Druie (Old Gaelic *dru*, 'oak', or perhaps Gaelic *druidh*, 'druid', 'magic-worker') before the bridge here was built.

Craigellachie (Moray) 'Rock of the stony place'. *Creag* (Gaelic) 'rock', 'crag', *ealeachaidh* 'stony', from Irish Gaelic *ailech* 'rock'.

Craigendoran (Argyll & Bute) 'Rock of the waters'. *Creag* (Gaelic) 'rock', *an t-* 'of the', *dobhráinn* 'waters'.

Craigentinny (Edinburgh) 'Rock of the fire', from Gaelic *creag* 'rock', *an* 'of', *teine* 'fire', indicating a signal point. 'Fox's rock', *an t-Sionnaich* (Gaelic) 'of the fox', has been suggested, but the Gaelic form is Creag an Teine.

Craiglockhart (Edinburgh) 'Camp rock'. *Creag* (Gaelic) 'rock', *luing* 'ship', *phort* 'landing-place' but used here as 'camping site'.

Craigmillar (Edinburgh) 'Rock of the bare height'. *Creag* (Gaelic) 'rock', *maol* 'bare, bald', *àrd* 'height'.

Craignure (Argyll & Bute) 'Yew tree rock'. *Creag* (Gaelic) 'rock', *an* 'of', *iubhair* 'yew tree'.

Crail (Fife) 'Boulder rock'. *Càrr* (Gaelic) 'rocky shelf', *ail* 'rock'. Recorded as Caraile in 1153.

Cramond (Edinburgh) 'Fort on the river Almond'. *Caer* (Brittonic) 'fort'; and see Almond. Recorded as Caramonth, 1178.

Crarae (Argyll & Bute) Perhaps 'Boggy place'. *Cràthrach* (Gaelic) 'boggy place', but the Gaelic form of the name is Carreibhe, suggesting *càrr* (Gaelic) 'rocky shelf', and perhaps *eighe* 'file'.

Crathie (Aberdeenshire) Perhaps 'shaking place', a reference to boggy ground. The Gaelic form is Craichidh, but an earlier Crathaigh from *crathach* 'shaking' has been postulated.

Crawford (South Lanarkshire) 'Crow ford': Scots *craw*, from Old English *crawe* 'crow'. Found as Crauford, c. 1150. 'Crooked ford' has also been proposed. Nearby Crawfordjohn adds 13th century John, son-in-law of the Sheriff of Lanark.

Cree, River (Dumfries & Galloway) 'Boundary', from *crìoch* (Gaelic); found as Crethe, 1301. Creetown is modern, 'town on the Cree'.

Crianlarich (Stirling) Probably 'House by the aspens': Gaelic *critheann* 'aspen tree', *laraich* 'house-site'. Noted as Creinlarach, 1603.

Crieff (Perth & Kinross) 'Place among the trees'. *Craobh* (Gaelic) 'tree'. Recorded as Creffe in 1178.

Crimond (Aberdeenshire) 'Hill-mound'. *Crech* (Gaelic, a form of *creachann*) 'hilltop', *monadh* 'hill', 'mountain'. Older forms are Creymund 1250; Crichmound 1550.

Crinan (Argyll & Bute) 'Place of the Creones' has been tentatively suggested: one of the western tribes identified by Ptolemy around CE150. The pre-Celtic root form is **cre-* and the Gaelic form is Crianan.

Cromarty (Highland) Originally 'crooked bay'. *Crom* (Gaelic) 'crooked', *bàgh* 'bay'. The modern Gaelic form is Crombà. Earlier forms are Crumbathyn in 1257 and Crombathie, 1296. Perhaps because of the existence of an alternative form incorporating *àrd* (Gaelic) 'height', the name altered to Cromardy in 1398, and Cromarte in 1565.

Cromdale (Moray) 'Bent haugh'. *Crom* (Gaelic) 'bent' with *dail* 'meadow', 'haugh'.

Crook of Devon (Perth & Kinross) 'Bend of Devon'. *Krókr* (Old Norse) 'crook,' 'bend'. The River Devon here takes a sharp turn westwards into Glen Devon. Crook of Alves in Moray and other Crooks have the same meaning.

Crossmyloof (Glasgow) Scotticised into 'Cross my palm' *(loof* 'palm'), it means 'Cross of Malduff', from *crois* (Gaelic) 'cross', and *Maolduibh* (Gaelic personal name) 'Malduff' (literally 'bald dark one').

Crossraguel (South Ayrshire) 'Cross of the bare fort.' 'Bare' probably means 'without a tower'. *Crois* (Gaelic) 'cross', *rathaig* 'small fort', *maol* 'bare'. Noted c. 1200 as Cosragmol, 1225 Crosragmol.

Croy (North Lanarkshire; Highland; South Ayrshire) 'Hard place'. *Cruadh* (Gaelic) 'hard'. Found as Croy in 1369.

Cruachan, Ben (Argyll & Bute) 'Heaped or haunched mountain'. *Cruach* (Gaelic) 'pile', 'stack': *cruachan* specifically means 'conical mountain'. Recorded as Crechanben c. 1375. Its name is often used without the prefix 'Ben', especially in Gaelic.

Cuillin (Highland) Possibly 'high rocks'. *Kiolen* (Old Norse) 'high rocks', 'ridge'. In Gaelic tradition the name was associated with the hero Cuchulainn, and 19th-century texts refer to the Cuchulin Hills. The Gaelic form is An Cuilfhionn or An Culthionn. A Norse origin is much more likely. The Rum mountains are also referred to as Cuillins.

Culbin (Moray) 'Back of the hill'. *Cul* (Gaelic) 'back', *bheinne* 'of the hill'. Noted as Coulbin, c. 1270.

Cullen (Moray) 'Holly'. *Culeann* (Gaelic) 'holly'. Noted as Inverculan, 1190; indicating that it was originally the stream name.

Culloden (Highland) Possibly, at the 'back of the little pool.' *Cul* (Gaelic) 'back', 'ridge', *lodan* 'little pool'. A document of 1238 records Cullodyn.

Culross (Fife) Probably 'holly wood'. *Culeann* (Gaelic) 'holly', *ros* 'wood'. Records show Culenros in 1110.

Cults (Aberdeen) It has been derived as 'the nooks', from Gaelic *cùiltean*, 'nooks', plural of *cùilt*. *Coillte* 'woods', has also been suggested. The terminal *-s* has been added in the Scots form to maintain the plural.

Culzean (South Ayrshire) 'Nook of birds'. *Cuil* (Gaelic) 'nook', 'corner', *ean* 'of birds'. Found in 1636 as Cullen.

Cumbernauld (North Lanarkshire) 'Meeting of the burns'. *Comar* (Gaelic) 'river confluence', *na* 'of the', *allt* 'stream', 'burn'. Recorded as Cumyrnald in 1417.

Cumbrae North Ayrshire) 'Island of the Cumbrians'. Old Norse adaptation of *Cymry* (Brittonic) tribal name of the Brittonic speakers, with *ey* 'island'. Recorded as Kumbrey 1270; Cumbraye 1330.

Cumnock (East Ayrshire) Perhaps 'crooked hill'. *Cam* (Gaelic) 'crooked', 'sloping', *cnoc* 'hill'. Recorded as Comnocke 1297; Cunnok 1461; Canknok 1548.

Cunningham (North Ayrshire) Of obscure but probably Celtic origin; attempts have been made to identify *cuinneag* (Gaelic) 'milk pail' in it. In 1153 it was recorded as Cunegan. It is also found in the form Cunninghame.

Cupar (Fife) Perhaps 'the confluence', from a lost Pictish word related to Gaelic *comar* 'confluence'. *Comphairt* (Gaelic) 'common', 'shared' with reference to pasture, is less likely.

Curly Wee (Dumfries & Galloway) This hill name as been derived as 'windy bend', from *cuir* (Gaelic) 'bend' *le* 'in the', *gaoith* 'wind', though Old Irish *cor* 'hill', has also been suggested.

Currie (Edinburgh) 'Boggy land'. *Currach* (Gaelic) 'bogland', 'marsh'.

Cursetter (Orkney) 'Cow pasture'. The first element is uncertain, with Old Norse *kyr* 'cow' suggested', with *saetr* 'farm', 'pasture land'.

Cushnie (Aberdeenshire) 'Cold', 'frosty', from *cuisneach* (Gaelic) 'frosty', 'freezing'. Perhaps an apotropaic name given to ward off the quality it describes. Cushnie near Alford has the helpful alternative name of 'Frosty Hill'.

D

Daer, River (South Lanarkshire) Perhaps from Irish Gaelic
   dér 'trickling', 'running'. Although the Clydes Burn,
   which joins it, is smaller, the combined stream bears the
   Brittonic name of Clyde.

Dairsie (Fife) The first element appears to be 'oak', from a
   lost Pictish word cognate with Gaelic *dair* or Brittonic
   derw. Recorded as Deruesyn, 1243: with a double suffix, -
   es and -*in*, found in other ancient Fife names. Also
   known as Osnaburgh, from an Old Norse personal name,
   with *burgh* (Scots, from Old Norse *borgr*, 'fort').

Dal- Placename prefix from Gaelic *dail* with a prime
   meaning of 'water-meadow', 'flood-plain'; also 'field'. A
   Brittonic-Pictish form *dol*, of the same meaning, gives
   some names in the former Pictland. Easily confused with
   Irish Gaelic *dál*, 'portion, tribe'.

Dalbeattie (Dumfries & Galloway) 'Meadow of the birch
   trees'. *Dail* (Gaelic) 'meadow', *beitheach* 'of birches'.
   Recorded as Dalbaty in 1469.

Dalgetty (Fife) 'Windy field'. *Dail* (Gaelic) 'field', *gaoithe*
   'of the winds'. Recorded as Dalgathyn, 1168.

Dalguise (Perth & Kinross) 'Haugh of fir'. *Dail* (Gaelic)
   'haugh, meadow', *giuthas* 'fir'.

Dalkeith (Midlothian) 'Field by the wood'. *Dol* (Brittonic)
   'field', and *coed* 'wood'. Early documents show Dalkied
   1140; Dolchet 1144; Dalketh 1145.

Dallas (Moray; Highland) 'Field by the waterfall'. *Dail*
   (Gaelic) 'field', *eas* 'waterfall'. Noted 1306 as Dolays.

Dalmally (Argyll & Bute) 'Site of Màillidh's church'. *Dail* (Gaelic) 'field', in this case specifically belonging to the church, *Màillidh* (Old Gaelic personal name). Màillidh's name is commemorated in a number of places.

Dalmarnock (Glasgow; Perth & Kinross) 'Site of Marnock's church'. *Dail* (Gaelic) 'field', in this case one specifically belonging to the church, *Mernóc* (Old Gaelic personal name). See Kilmarnock.

Dalmellington (South Ayrshire). Perhaps 'hill-fort place'. The oldest records show Almelidun, either from Brittonic *al* or Gaelic *ail* 'rock', with dun 'hill', 'hill fort'. The prefix had changed to Dal- and the suffix to -ton by the 13th century. The middle element may relate to Gaelic *meallan* 'mounds'.

Dalmeny (Edinburgh) Perhaps 'My Ethne's meadow'. *Dail* (Gaelic) 'meadow', *mo* 'my', *Eithne* (Old Gaelic personal name). As the mother of St Columba, Ethne was a revered figure. But older forms include Dumanie (c. 1180) and Dunmany (1296), suggesting Gaelic *dùn*, 'fort', from Brittonic *din*, perhaps with Brittonic *meini*, 'stones'. See also Kilmany.

Dalmuir (West Dunbartonshire) 'The big field'. *Dail* (Gaelic) 'field', *mòr* 'big'. Noted c. 1200 as Dalmore, but the suffix later became confused with *muir* (Scots) 'moor'. Found as Dalmuire in 1680.

Dalnaspidal (Perth & Kinross) 'Field of the refuge'. *Dail* (Gaelic) 'field', *na* 'of' the', *spideal* 'refuge', 'hospice'.

Dalry (Edinburgh; Dumfries & Galloway; North Ayrshire) Probably 'field of the heather', either from Gaelic *dail* or Brittonic *dol* 'field', with *fhraoich* or *wrūg* 'heather'; or the latter part may be from Gaelic *ruigh* 'slope'. It has also been suggested that the D&G name, in full St John's Town of Dalry, could be derived as 'the king's meadow': *righ* (Gaelic) or *rūy* Brittonic) 'king'. Recorded as Dalrye in 1497.

Dalserf (South Lanarkshire) 'Place of St Serf's church'. *Dol* (Brittonic) 'field', in this case specifically belonging to the church'; *Serf* (personal name from Latin *servus*, 'slave').

Dalrymple (South Ayrshire) 'Field of the winding pool'. *Dail* (Gaelic) 'field', *crom* 'bent', 'winding', *poll* 'pool'. The *c* of *crom* has been lost. Found in 1300 as Dalrimpill.

Dalwhinnie (Highland) Apparently the 'field of the champion'. *Dail* (Gaelic) 'field', *cuingid* 'champion'. This may refer to some historic or legendary contest.

Dalziel (North Lanarkshire) 'White meadow'. *Dail* (Gaelic) 'field, meadow', *gheal* 'white'. There are numerous Dalziels and Dalzells throughout the country. The *z*, as in many other cases, is actually a form of written *gh*. Recorded in 1200 as Dalyell.

Darvel (East Ayrshire) Perhaps 'stream by the township'. *Dobhar* (Gaelic) 'water', *bhaile* 'of the township'. Noted on Blaeu's map of Scotland as Darnevaill.

Dava (Highland) 'Ford of the stags, or oxen'. Gaelic An Damháth, *An* 'the', *damh* 'ox', 'stag', *àth* 'ford'.

Daviot (Highland; Aberdeenshire) Perhaps from an ancient tribal name, latinised as *Demetae*. Its Gaelic form is Deimhidh, from a conjectured root form **dem* (Pictish-Brittonic) 'fixed,' 'sure'. It is cognate with Welsh Dyfed. Noted as Dauyot, 1136.

Dean, River (Midlothian) 'Valley', 'den'. *Denu* (Old English) 'dene'. Found c. 1145 as Dene.

Dearg, Ben (Highland; Aberdeenshire) 'Red mountain'. *Beinn* (Gaelic) 'mountain', *dearg* 'red'. There are numerous mountains of this name throughout the Highlands, and also Càrn Deargs.

Dee, River and Strath (Aberdeenshire; Dumfries & Galloway) This river-name has a complex history, not yet fully explored. Its Gaelic name *Dé* has been related to *Dia*, 'god', though the river's gender is feminine. Its early-Celtic root is *Deua*, meaning a female divinity; it shares this with the Don. The Galloway Dee is also known as the Black Water of Dee.

Deer (Aberdeenshire; Dumfries & Galloway) 'Forest grove'. *Doire* (Gaelic) 'grove', normally of oaks, from Old Irish *daur*, 'oak', and noted in the eponymous Book of Deer as Dear, c. 1150.

Denny (Falkirk) 'Valley'. *Denu* (Old English) 'valley'. Noted as Litill Dany in 1510. Nearby is Dennyloanhead, 'valley at the head of the lane'.

Deveron, River (Aberdeenshire) 'Black Earn'. This river was originally called *Eron*, perhaps from Old Irish *Erin*, or more probably from an older pre-Celtic source (see Earn). *Dubh* (Gaelic) 'dark' is a later prefix, perhaps to distinguish this river from the Findhorn, as with the Adder rivers in the borders, and the various Esks. It was recorded as Douern in 1273.

Devon River and **Glen** (Perth & Kinross) 'Black stream'. *Dub* (Old Irish) 'black'; **ona* (pre-Celtic) suffix indicating 'water', 'river'. It has also been suggested that it may stem from an inferred Brittonic word **domnona*, 'deep or mysterious one'. Its valley was recorded as Glendovan around 1210.

Diabaig (Highland) 'Deep bay', from *djúp* (Old Norse) 'deep', and *vik* 'bay'.

Dingwall (Highland) 'Parliament field'. *Thing* (Old Norse) 'parliament assembly', *vollr* 'field', 'open space'. Exact parallels to the name are found in other areas of Norse influence (e.g. Tingwall on Shetland, and Tinwald in Dumfries & Galloway). Recorded as Dingwell, 1227.

Dinnet (Aberdeenshire) 'Place of shelter. *Dìon* (Gaelic) 'shelter', *ait* 'place'.

Dochart, River and **Glen** (Stirling) 'Evil scourer'; the river-name, from *do-* (Gaelic prefix) 'evil', and *cartaim*, 'I cleanse, scour'. Noted c. 1200 as Glendochard.

Docharty, Glen (Highland) This may have the same derivation as Dochart; its Gaelic form is Gleann Dochartaich. *Dochair* in current Gaelic means 'hurt, misery, pain'.

Dochfour (Highland) 'Pasture area'. *Dabhach* (Gaelic) 'field measure', *phùir* 'pasture'.

Doll, Glen (Angus) 'Glen of the meadows', from Pictish-Brittonic *dol*, 'meadow', 'valley'.

Dollar (Clackmannanshire) Place by the 'ploughed field'. Dol (Pictish-Brittonic) 'field', *ar* 'arable', 'ploughed'. Noted in the *Pictish Chronicle*, c. 970, as Dolair and in 1461 as Doler.

Dolphinton (Midlothian) 'Dolfin's place'. A charter of 1253 records this as Dolfinston; Dolfin was the brother of the first earl of Dunbar.

Don, River (Aberdeenshire) The Gaelic name is Deathan, from the same root as the Dee: **deua* (Celtic) 'god', with suffix **ona* indicating 'water', 'river'. The form Don is found from 1170. Belief in a river-spirit is indicated. See Dee.

Donibristle (Fife) 'Breasal's fort'. *Dùnadh* (Gaelic) 'camp'; *Breasail* (Old Irish personal name). Noted as Donibrysell, c. 1165.

Doon, River, Loch (South Ayrshire) The same name as Don, from the Celtic root forms, **deu* and **ona* 'river goddess'; noted as Don, 1197.

Dorain, Ben (Argyll & Bute) 'Hill of the streamlets'. *Dobhráinn* (Gaelic) 'of streams'. There is also Beinn Dhorain above Strath Kildonan, in Sutherland, though this has also been derived from *dòbhran* (Gaelic) 'otters'.

Dores (Highland) 'Black woods'. *Dubh* (Gaelic) 'black', and *ros* 'wood'. Recorded as Durris, 1263 (see Durrisdeer).

Dornie (Highland) 'Place of pebbles'. *Dornach* (Gaelic) 'with pebbles'.

Dornoch (Highland) 'Place of pebbles'. *Dornach* (Gaelic) 'with pebbles'. The root-word *dorn* means 'fist'. The name was recorded as Durnach in 1150. Dornock on the Solway coast is probably from a related Brittonic form.

Douglas (Dumfries & Galloway; South Lanarkshire) 'Dark water'. *Dubh* (Gaelic) 'black', *glas* is one of several Gaelic 'stream' words. Recorded as Duuelglas around 1150. The name is also found in other localities in Argyll and Angus; some, like Castle Douglas and Douglastown, are later creations taken from the family name rather than any water feature.

Doune (Stirling) 'Castle, fortified place'. *Dùn* (Gaelic) 'fort'. The local names Dounie, Downie, in various districts, have the same derivation.

Dounreay (Highland) Possibly 'fortified rath'. *Dùn* (Gaelic) 'fortified place', *rath* 'circular fort' or 'broch'.

Dowally (Perth & Kinross) 'Black cliff'. *Dubh* (Gaelic) 'black', *àille* 'cliff'. Noted as Dowalye in 1505.

Dreghorn (Edinburgh; North Ayrshire) 'House on the river-spit', from Old English *draeg* 'river-spit' and *erne* 'house', 'hut'. Recorded as Dregerne c1240.

Drem (East Lothian) 'Ridge'. *Druim* (Gaelic) 'ridge', 'hump'.

Drumalban 'The ridge of Scotland'. The great spinal watershed that runs up from central to northern Scotland, from *druim* (Gaelic) 'ridge'; and *Albainn* 'of Scotland'.

Drumchapel (Glasgow) 'Ridge of the horse'. *Druim* (Gaelic) 'ridge', *chapuill* 'of the horse'.

Drumbeg (Highland) 'Little ridge'. *Druim* (Gaelic) 'ridge', 'hump', *beag* 'small'.

Drumclog (South Lanarkshire) 'Ridge of the rock' or of 'the bell'. *Druim* (Gaelic) 'ridge', 'hump'; *clog* may be Brittonic 'rock, crag', or Gaelic *clag* 'bell'.

Drumnadrochit (Highland) 'Ridge of the bridge'. *Druim* (Gaelic) 'ridge', *na* 'of the'; *drochaid* 'bridge'.

Drunkie, Loch (Stirling) 'Loch of the litle ridge'. *Loch* (Gaelic) 'lake, loch', *dronnaig* 'little ridge', 'knoll'. An alternative is Irish Gaelic *drong* 'meeting' with *-aidh* suffix for 'place'.

Dryburgh (Borders) 'Dry place', perhaps immune to flooding from the nearby Tweed. Old English *dryge* 'dry', *burh* 'town'. Early records show Drieburh 1160, Dryburg 1211.

Dryhope (Borders) Probably 'dry hollow'. Old English *dryge* 'dry' and *hop* 'enclosed valley'.

Drymen (Stirling) 'On the ridge'. *Drumein* (Gaelic dative/locative of *druim*) 'on the ridge'. Noted in 1238 as Drumyn.

Duart (Argyll & Bute) 'Black point'. *Dubh* (Gaelic) 'black', *àird* 'point', 'height'.

Duddingston (Edinburgh) 'Dodin's farmstead'. *Dodin* (Old English personal name), *tūn* 'farm, village', noted as Dodinestun, 1150. Some time prior to that it had a Brittonic name, Trauerlen, from *tref* 'place, settlement', *yr* 'of the', *llin* 'lake'.

Dufftown (Moray) Town founded in 1817 by James Duff (from Gaelic *dubh* 'black') the fourth earl of Fife, after whom it is named.

Duich, Loch (Highland) 'Duthac's loch'. *Loch* (Gaelic) 'lake', 'loch', *Dubhthaich* (Gaelic personal name) 'of Duthac', replacing an earlier lost name. St Duthac (11th century) was a venerated figure; see Black Isle, Tain.

Duirinish (Highland) 'Deer ness'. *Dyr* (Old Norse) 'deer', *nes* 'point', 'headland'; recorded in 1567 as Durynthas. *See* Durness.

Dull (Perth & Kinross) 'Field', 'haugh'. *Dol* (Pictish-Brittonic) 'meadow'.

Dullatur (North Lanarkshire) 'Dark slopes'. *Dubh* (Gaelic) 'black', *leitir* 'hillside'. It is noted on Blaeu's map as Dulettyr.

Dulnain, River (Highland) 'Flood stream'. *Tuil* (Gaelic) 'flood', with *-ean* suffix indicating a stream. The Gaelic name is Abhainn Tuilnean.

Dumbarton (West Dunbartonshire) 'Stronghold of the Britons'. *Dùn* (Gaelic) 'fortified stronghold'; *Breatainn* 'Britons'. Records show Dunbretane from 1300. The Britons themselves called it Alcluith, 'rock of Clyde'. The name of its county, Dunbartonshire, shows a more 'correct' form, though the change back from *m* to *n* was made in modern times.

Dumfries (Dumfries & Galloway) 'Fortress of the woodland'. *Dùn* (Gaelic) 'fortified stronghold', perhaps replacing an earlier Brittonic *din*, *phris* (Gaelic genitive of *preas*) 'of the woodland copse'. Found as Dunfres, c. 1183.

Dumyat (Stirling) 'Hill of the Maeatae'. *Dùn* (Gaelic) originally indicated a hill, but so many hilltop sites were fortified that it also acquired the sense of 'hill-fort'. The name incorporates that of a tribal group identified by the Romans in CE208. Myot Hill, not far away, also appears to preserve the name.

Dunadd (Argyll & Bute) 'Fortress of the Add'. *Dùn* (Gaelic) 'fort'. The etymology of *Add* is unclear, though it presumably is from the adjacent River Add, and may be pre-Celtic. The oldest recorded spelling is Duin Att, 683.

Dunbar (East Lothian) 'Fort on the height'. *Din* (Brittonic) 'fort', *barr* 'height'. Recorded 709 as Dynbaer: the later *dun-* form of the prefix has been ascribed to Anglian or Gaelic influence.

Dunbeath (Highland) 'Hill of birches'. *Dùn* (Gaelic) 'hill, mound', *beith* 'birch tree'. Noted in this form 1450.

Dunblane (Stirling) 'Field of St Blane'. *Dùn* (Gaelic) is 'fort' but in the 9th century Dulblaan is noted, with Pictish-Brittonic *dol* 'field'; *Bláán* (Old Irish personal name) Blane, 6th century missionary. By c.1200 Dumblann is recorded.

Duncansby Head (Highland) 'Cape of Dungal's place'. *Dungal* (Gaelic personal name); *by* (Old Norse) 'village'. Noted in the *Orkneyinga Saga*, c. 1225, as Dungalsbaer.

Dundee Commonly derived as the 'fort of Daig'. *Dùn* (Gaelic) 'fortified place'; *Daig* (personal name) of unknown connection. Other possible derivations include *Dun-dubh* (Gaelic) 'dark hill', or *Dun-Dè* (Gaelic genitive of *Dia*) 'hill of God'. Early records include Donde 1177; Dunde 1199; Dundho, Dundo 1200.

Dundurn (Perth & Kinross) 'Fort of the fist'. *Dùn* (Gaelic) 'fort', *dorn* 'fist'. Noted in 603 as Duin Duirn, it was a Pictish stronghold.

Dunfermline (Fife). The first element is *dùn* (Gaelic) 'hill' or 'fort', the latter elements are of uncertain derivation, though local stream names *Ferm* and *Lyne*, of Pictish origin, have been noted. Records show Dumfermelyn 1100; Dumferlin 1124; Dunferlyne 1375.

Dunino (Fife) 'fort of the meeting-place', from Gaelic *dùn*, 'hill' and *aonach*, which as well as 'ridge,' 'open moor', can mean 'place of assembly'. Noted as Duneynach, 1250.

Dunkeld (Perth & Kinross) 'Fort of the Caledonians'. *Dùn* (Gaelic) 'fort', *Chailleainn* 'Caledonians' – the Picts had a stronghold here in the first millennium CE. Early records reveal Duincaillen 865; Dun-calden and Dunicallenn c. 1000.

Dunlop (Dumfries & Galloway) 'Fort of the bend'. *Dùn* (Gaelic) 'fort', *luib* 'bend'.

Dunnet (Highland) 'Fort'. The prefix is *Dùn* (Gaelic) 'fort' possibly replacing Pictish *din*, with -*aid* suffix of unclear meaning. An Old Norse name might be expected, particularly for Dunnet 'Head'. Ptolemy noted it as *Tarvedum* 'bull place' – see Thurso.

Dunnichen (Angus) 'Nechtan's fort'. *Dùn* (Gaelic) 'fort'; perhaps replacing Pictish *din*; *Nechtan* (gaelicised form of a Pictish personal name, Nehton). Probably named for the early 7th century Pictish King Nechtan, and noted as Duin Nechtain in the *Annals of Tighernach* (11thc).

Dunoon (Argyll & Bute) 'Fort by the river'. *Dùn* (Gaelic) 'fort', *obhainn* (adjectival variant of *abh*) 'river'. Recorded as Dunhoven 1270; Dunnovane 1476.

Dunottar (Aberdeenshire) 'Fort on the shelving ground'. *Dùn* (Gaelic) 'fort', *faithir* 'shelved or terraced slope'. The oldest reference is Duin-foither, 681.

Dunphail (Moray) 'Fort with the palisade'. *Dùn* (Gaelic) 'fort', *fàl* 'hedge', 'palisade'. Noted c. 1250 as Dunfel. The indication is an early fortification of wood rather than of stone.

Dunragit (Dumfries & Galloway) 'Fort of Rheged'. *Dùn* (Gaelic) 'fort' replacing earlier Brittonic *din*. This was a centre of Rheged, for a time part of the kingdom of Strathclyde which stretched from Dumbarton to Westmorland.

Dunrossness (Shetland) 'Cape of the roaring whirlpool'. *Dynr* (Old Norse) 'loud noise', *röst* 'whirlpool', *nes* 'cape', 'point'. There are numerous tidal races called Roosts in the Northern Isles.

Duns (Borders) 'Fortified hill'. It may be from *dinas* (Brittonic), *dùn* (Gaelic) or Old English *dān*, all meaning 'fortified hill'. The *s* is a later addition, possibly meant as a plural, found in this form in 1296.

Dunsinane (Perth & Kinross) Perhaps 'hill of the paps, or nipples'. *Dùn* (Gaelic) 'hill', *sineachan* 'nipples'. Recorded in the *Pictish Chronicle* c. 970, as Dunsinoen.

Duntocher (West Dunbartonshire) 'Causeway fort'. *Dùn* (Gaelic) 'fort', *tóchar* 'causeway', 'road'. It was a Roman road, by the end of the Antonine Wall.

Durness (Highland) 'Headland of the deer'. *Dyr* (Old Norse) 'deer', *nes* 'headland'. Noted c. 1230 as Dyrnes.

Duror (Highland) 'Hard water'. A river name, from *dur* (Gaelic) 'stiff', 'hard', *dobhar*, 'water'. An old form is Durdowar.

Durrisdeer (Aberdeenshire) 'Entrance to the forest'. *Dorus* (Gaelic) 'entrance', *doire* 'forest'. Noted in 1275 as Durisdeir. But the south Deeside location and forest of Durris is likely to be from *dubh* (Gaelic) 'black' and *ros* (Old Gaelic) 'wood'.

Dyce (Aberdeenshire) Possibly 'southwards'. *Deis* (Gaelic locative of *deas*) 'to the south'. This may have been a reference to the location. Old Norse *dys* 'cairn' has also been suggested.

Dysart (Fife) 'Hermit's place'. *Diseart* (Gaelic) 'hermit's place'. Recorded as Disard, c. 1210. The hermit was St Serf who lived for a time in a cave here.

E

E, River (Highland) A candidate for the shortest name in Britain, this stream joins the Fechlin (Gaelic *fiach* 'worthy', *linne* 'waterfall, pool'), above Foyers on the east side of Loch Ness. It stems from Old Norse *àa*, 'water', cognate with Old English *ea*, as in Eye.

Eaglesham (East Renfrewshire) Possibly 'church village'. *Eaglais* (Gaelic) 'church'; *ham* (Old English) 'village'. Recorded as Egilshame 1158, Eglishame 1309.

Earlsferry (Fife) Established in the 12th century as a ferry point across the Firth of Forth. Its name suggests that by the later 1100s Gaelic names were no longer being given in Fife.

Earlston (Borders) Possibly 'Earcil's hill'. *Earcil* (personal name); *dun* (Old English) 'hill'. Early records show Ercheldon 1144; Ercildune 1180.

Earn, River, Strath and **Loch** (Perth & Kinross) Traditionally derived as 'Erin'. The name *Eireann* (Old Irish Gaelic) 'of Erin', has been taken to mark the eastward expansion of the Scots from Dàl Riada in the 6th and 7th centuries, awarding familiar names to new landscapes. *Erin* is an ancient Gaelic name linking a mythical goddess-queen with the land itself. Possibly goddess and river (Irish river and lough Erne) both stem from an early- or pre-Celtic root-form *ar-* indicating flowing water.

Eassie (Angus; other regions) 'Water'. *Eas* (Gaelic) 'water', 'waterfall'; the *-ie* ending, corresponding to Gaelic *-aidh*, is often found in river-names in regions where Pictish was spoken. A frequent local name, sometimes found as Essie or Essy. But Essich, near Inverness, appears to show the Gaelic locative suffix *-ach*: 'place by the water'.

East Kilbride (South Lanarkshire) 'Church of (St) Bride'. *Cill* (Gaelic) 'church', *Brigid* name of a saint who took on attributes of the Celtic goddess Brid. The old village here was recorded as Kellebride in 1180. 'East' was added later to distinguish it from West Kilbride, thirty miles away.

East Linton (East Lothian) 'Flax enclosure'. *Lin* (Old English) 'flax'; *tūn* (Old English) 'village'. It was recorded as Lintun in 1127. 'East' was added later to distinguish it from West Linton, thirty miles south-west.

Ecclefechan (Dumfries & Galloway) 'Fechin's church' from *eaglais* (Gaelic) 'church'; *Fechin* (Irish Gaelic personal name), the 7th-century Irish missionary. An alternative derivation is the Brittonic *egles-bychan*, 'little church'. Noted as Eglesfeghan, 1303.

Eccles (Dumfries & Galloway) 'Church', from Brittonic *egles*, derived from Latin *ecclesia*.

Eck, Loch (Argyll & Bute) 'Horse loch'. *Loch* (Gaelic) 'lake', 'loch', *each* 'horse'. The horse may have been a water kelpie, or *each uisge*.

Eday (Orkney) 'Island of the isthmus'. *Eidh* (Old Norse) 'isthmus', and *ey* 'island'.

Edderton (Highland) 'Place between mounds'. *Eadar* (Gaelic) 'between', and *dùn* 'mound', 'hillock'. Recorded as Ederthayn in 1225.

Eddleston (Borders) 'Edulf's place', from the late 12th century. *Edulf* (Old English personal name) and *tūn* 'farmstead'. Previously recorded as first Penteiacob (Brittonic) 'Headland of James's town'; then Gillemorestun, from *Gillemor* (Gaelic) 'Gilmour' and *tūn* – indicating three proprietors from three different language groups before the name became fixed, with Edoluestone found c. 1200.

Eddrachillis (Highland) 'Place between two kyles'. *Eadar* (Gaelic) 'between', *da* 'two', *chaolais* 'kyles', 'narrows'.

Eden, River (Fife; Borders) This river-name, also found in Cumberland and Kent in England, is of uncertain origin. The Cumbrian Eden is *Ituna* on Ptolemy's map, and has been traced back to a conjectural Primitive Welsh **idon*.

Edinburgh 'Edin-' remains unexplained; *eiddyn* (Brittonic) 'rock face', with *burh* (Old English) 'stronghold' is no longer considered likely. The 'din' part is *din* (Britonnic) 'stronghold' or 'fort'. The earliest reference, CE638, is simply Etin. Dùn Eideann remains the Gaelic name for the city.

Ednam (Borders) 'Place on the Eden', recorded as Aedenaham, c. 1105; Hedenham, 1316. *See* Eden.

Edrom (Borders) 'Township on the Adder'. One of the few Scottish places with the Old English *-ham* suffix, which seems to have become obsolete around the time the Anglians were establishing themselves in the south-east. The first element is the river-name Adder.

Edzell (Angus) Known as 'Aigle' in Scots, 'Eigill' in Gaelic, with the 'z' representing 'gh', the source of the name is unclear. It is found in 1250 as Adel. The name was transferred from a nearby castle site when the town was established in 1818.

Egilsay (Orkney; Shetland) Possibly 'church island'. *Eaglais* (Gaelic) 'church'; *ey* (Old Norse) 'island'. Noted as Egilsey in the *Orkneyinga Saga*, c. 1225. The hybrid formation is unusual, especially here, and 'Egil's (Old Norse personal name) isle' has been suggested for both islands.

Eglinton (North Ayrshire) Farm of Aegel's folk. *Aegel* (Old English personal name), *ing* 'of the people', *tūn* 'farmstead'. Noted as Eglunstone, 1205.

Eigg (Highland) Probably '(island with) the notch'. *Eag* (Gaelic) 'notch', 'nick', 'gap'. A wide rift or notch runs through the island from southeast to northwest. Recorded as Egge 1292, Egg 1654.

Eildon Hills (Borders) Called Trimontium by the Romans; may be from Old English *aeled* 'fire', *dūn* 'hill'. The form Aeldona is found around 1120.

Eishort, Loch (Highland) 'Isthmus firth', from *eidh* (Old Norse) 'isthmus', and *fjordr* 'firth', with Gaelic *loch* a later addition.

Elderslie (Renfrewshire) 'Alder lea'. *Elloern* (Old English) 'alder', *li* 'meadow'. Found in 1398 as Eldersly.

Elgin (Moray) Perhaps 'Little Ireland'. *Ealg* (Gaelic) early name for Ireland, but also, perhaps by extension, 'noble, excellent', *-in* (Gaelic diminutive suffix) 'little'. Found in its present form in 1140, it may simply mean 'worthy place'. Eilginn is the Gaelic name. Compare Blair Atholl and Glenelg.

Elgol (Highland) 'Fold of the Stranger', has been put forward as the source: *Fàl* (Gaelic) 'fold', and *a' ghoill* 'of the Gall, or stranger'.

Elie (Fife) The form seems close to *eilean* (Gaelic) 'island', and the present harbour was an island until the late 18th century; but 'place of the tomb' has also been proposed, from *ealadh* (Gaelic) 'tomb'; or *ayle* (Scots) 'covered cemetery'. Gaelic *éaladh* 'creeping' can also indicate a narrow passage for boats. It is recorded as Elye 1491, The Alie c. 1600.

Elliot (Angus) 'Mound'. This stream-name may stem from Gaelic *eileach*, 'mill-dam', 'weir', 'mound'; and be cognate with Elliock in Dumfries & Galloway.

Ellon (Aberdeenshire) Possibly 'green plain or meadow', from *àilean* (Gaelic) 'green place', meadow'. This derivation suits the location; though *eilean* (Gaelic) 'island', is also appropriate. Noted in the *Book of Deer*, c. 1150, as Helian and Eilan, which suggest the latter form.

Elphinstone (East Lothian) Perhaps 'Alpin's fort'; from Gaelic *Alpin* (personal name) and *dún*, 'fort'. Found as Elfyngston, around 1320. Perhaps under Anglian influence, some suffixes in the south-east have been converted from -dun to -ton. *See* Earlston. Port Elphinstone in Aberdeenshire is 19th century, from the landowner Sir Robert Elphinstone.

Elrick (Highland; Aberdeenshire) This name is found in a number of localities. It is cognate with Elrig (Dumfries & Galloway) and is also found as Eldrick. A likely source is Gaelic *eilerg*, derived by metathesis of *r* and *l* from Old Irish Gaelic *erelc*, 'ambush', with the sense of 'deer trap': a cul de sac into which hunted deer were driven for slaughter.

Embo (Highland) 'Eyvind's steading'. *Eyvin* (Old Norse personal name), *ból* (Old Norse, shortened form of *bolstadr*) 'farmstead'. Noted as Ethenboll, c. 1230.

Eoropaidh (Western Isles) 'Beach village'. A gaelicised form of Old Norse *eyrar-by* 'shore settlement'.

Erbusaig (Highland) 'Erp's bay'. *Erp* (Old Norse personal name; *-aig* a Gaelic form of *vik* (Old Norse) 'bay'.

Eriboll (Highland) 'Farm on the ridge'. *Eyri* (Old Norse) 'tongue of land', *ból* (from *bolstadr*) 'farmstead'. It has been suggested that this abbreviation here and elsewhere shows Gaelic taking over from Old Norse . The Gaelic name is Euraboll. In 1499 it is noted as Erribull.

Ericht, River, Loch (Highland; Perth & Kinross) Perhaps 'beauteous', from *eireachdas* (Gaelic) 'beauteous'. The Erichdie Water in Atholl may be related though it has also been linked to *eireachda* (Gaelic) 'assembly'.

Eriskay (Western Isles) 'Eric's island'. *Erik* (Old Norse personal name), *ey* 'island'. Noted as Yriskay, 1558. The Gaelic form is Eiriosgaigh.

Errol (Perth & Kinross) Tentatively linked to *ar ole* (Brittonic–Pictish) 'on the ravine', but there is no ravine here. Perhaps the first element is Gaelic *éarr* 'boundary'. Noted as Erolyn, c. 1190. Port Errol on Cruden Bay, Aberdeenshire, is a modern name.

Erskine (Renfrewshire) Possibly 'high marsh', from *àrd* (Gaelic) 'high', and *sescenn* 'marsh'; but Brittonic *ir ysgyn*, 'green ascent' has also been put forward. Noted as Erskin, 1225.

Esk, River (Angus; Dumfries & Galloway; East Lothian) 'Water', a basic river-name. *Uisge* (Gaelic) 'water'; the numerous Esks testify to its currency. From a pre- or early-Celtic root *esc*; noted as Esce (Lothian), 800; Esche (Angus), 1369. It appears to be a preferred name for rivers reaching the sea close to each other, like the North and South Esks.

Etive, River and **Loch** (Argyll & Blue) The Gaelic name is Eitche, which has been taken to be from *Eitig* (Old Irish feminine personal name) 'foul one', indicating a malevolent spirit, inspired perhaps by the turbulent tidal entrance. A connection with *èiteag* (Gaelic) 'white pebble', 'quartz', has also been suggested. A pre-10th-century source has Loch-n-Eite.

Ettrick (Borders; Argyll & Bute) The name is that of the river, the Ettrick Water, also applied to Ettrick Forest, Ettrick Pen (the latter a Brittonic word) Ettrickbridge, etc The first part may be related to *eadar* (Gaelic) 'between'. Another suggestion is *atre* (Brittonic) 'playful', as a description of the river here. The origin remains uncertain, and it may be from a pre-Celtic root. The forms Ethric and Etryk are recorded from around 1235.

Evanton (Highland) 'Evan's town'. The village was founded around 1810 by the landowner Evan Fraser of Balconie.

Eyemouth (Borders) 'At the mouth of the Eye Water'. The river-name is tautologically derived from *ea* (Old English) 'river'; 'water' was added when the old meaning was lost.

Eynhallow (Orkney) 'Holy isle'. In the *Orkneyinga Saga* of c. 1225, it is recorded as Eyin Helga, from Old Norse *ey* 'island', and *heilag-r* 'saint'.

F

Fair Isle (Shetland) 'Sheep Island'. *Faer* (Old Norse) 'sheep'; isle (later English translation of *ey* 'island'). A record of 1529 shows Faray.

Falkirk 'Speckled church'. *Fawe* (Scots) 'speckled'; *kirke* (Scots) 'church' from Old English *cirice*. Most Gaelic names were maintained in an approximation of their original form, with a Scots pronunciation, but this is a Scots translation of the original form, Egglesbreth, from Gaelic *eaglais* 'church', and *breac* 'speckled', first recorded in the early 12th century. It is presumed that a church here was built of variegated stone. It was recorded as Faukirke in 1298, Falkirk from 1458.

Falkland (Fife) The origin remains uncertain. Possible associations with falconry have been made from *falca* (Old English) 'falcon', while -land may be from Pictish or Gaelic *lan* or *lann* 'enclosure'. It was a royal hunting estate. Early records show Falleland 1128, Falkland 1150, Falecklen 1165. The original settlement name was Kilgour, 'Gabran's church'.

Falloch, Glen (Stirling) Possibly 'Glen of hiding'. *Gleann* (Gaelic) 'valley', and *falach* 'place of concealment', though a connection with Irish Gaelic *fail* 'ring', has also been suggested.

Fannich, River, Loch and **Hills** (Highland) *Fàn* (Gaelic) indicates a gentle slope, and *Fanaich* may indicate the dip slopes, important for summer grazing. A Pictish water-name has also been suggested, cognate with the Welsh verb *gwanegu*, 'to rise in waves', perhaps from a continental Celtic root-form *ven or *van.

Fare, Hill of (Aberdeenshire) 'Watch hill'. *Faire* (Gaelic) 'watchfulness', 'sentinel', with 'hill' substituted for Gaelic *cnoc* in the post-Gaelic era.

Farg, Glen (Perth & Kinross) 'Ferocious glen'. *Gleann* (Gaelic) 'valley', and *fearg* 'anger'.

Farrar, River and **Strath** (Highland) First recorded on Ptolemy's map of Scotland, c. CE 150, as *Varar*, this much-discussed name hints for some at a non-Celtic Indo-European language spoken in the territory of the Picts. The upper reaches of the river have become called Glen Strathfarrar, perhaps indicating uncertainty as to how the valley should be classed.

Faskally (Perth & Kinross; Highland) 'Stance by the ferry'. *Fas* (Gaelic) in placenames has the sense of 'stance', *calaidh* 'ferry'. Noted in 1611 as Faschailye. The *Fas*-element in many local placenames indicates the extent of sheep and cattle-droving across the country to markets or new grazings.

Faslane (Argyll & Bute) 'Stance on the enclosed land'. *Fas* (Gaelic) 'stance', *lainne* (Gaelic locative form of *lann*) 'enclosed ground', 'field'. Found in this form in 1531.

Fassifern (Highland) 'Stance of the alder trees'. *Fas* (Gaelic) 'stance', *feàrna* 'of the alders'. Also found as Fassfern. Noted as Faschefarne, 1553.

Fauldhouse (West Lothian) 'House on the fallow land'. *Falh* (Old English) 'fallow land,' *hus* 'house'.

Fearn (Highland) 'Place of the alders'. *Feàrna* (Gaelic) 'alder'. The name was transferred to this location when the monastery at West Fearn, near Edderton, was refounded here as an abbey in 1227. Noted in 762 as Ferna; 1529 as Ferne. In Gaelic it was known as Manachainn Rois, 'monastery of Ross'. Fearnan by Loch Tay has the same derivation.

Feshie, River and **Glen** (Highland) 'Boggy meadowland'. *Féith* (Gaelic) 'boggy place', *-isidh* Gaelic suffix denoting pasture-land, derived from *innse*, meaning 'meadow' as well as 'island'.

Fetlar (Shetland) Perhaps from Old Norse *fetill*, 'belt', but a lost Pictish origin has also been suggested.

Fettercairn (Aberdeenshire) 'Wooded slope'. *Faithir* (Gaelic) 'terraced slope'; *cardden* (Pictish) 'thicket'. Recorded in the *Pictish Chronicle* around 970 as Fotherkern.

Fetteresso (Aberdeenshire) 'Watery slope'. *Faithir* (Gaelic) 'terraced slope', *easach* 'watery'. Recorded as Fodresach in the *Pictish Chronicle*.

Fiddich, River and **Glen** (Moray) Fidach was one of the ancient province names of Pictland and it seems likely that this name preserves it. The root element *fid* is likely to be from a personal name.

Fife Attributed in legend to the personal name Fib, a legendary precursor of the Picts, one of the seven sons of Cruithne who gave their names to the provinces of Pictland. Recorded as Fib 1150, Fif 1165. The Gaelic form is Fiobha.

Findhorn (Moray) 'White water'. *Fionn* (Gaelic) 'white'; **eren* (pre-Celtic river-name, *see* Earn). In Gaelic it is Uisge Eire. This name has sometimes been seen as a form of Irish Gaelic *Erin*. Noted in 1595 as Fyndorn. The village takes its name from the river.

Findochty (Moray) 'House on the fair or bright land-measure'. *Fionn* (Gaelic) 'fair', 'white', *dabhach* 'land measure', *taigh* 'house'. Noted in 1440 as Fyndectifeilde; the Scots 'field' has been lost.

Finlaggan (Argyll & Bute) 'Fair hollow'; Gaelic *fionn*, 'fair,' 'bright', and *lagan*, 'hollow'. Found as Finlagan, 1427. The site, headquarters of the lords of the Isles in the 15th century, has also been identified with the 6th-century St Findlugan.

Finnan, River and **Glen** (Highland) 'Fingon's Glen'. *Gleann* (Gaelic) 'glen', *Fhionghuin* (Gaelic personal name).

Finnart (Argyll & Bute, Galloway) 'Bright height'. *Fionn* (Gaelic) 'fair', 'bright', *àird* 'height'.

Finnieston (Glasgow) This name was given to the Glasgow district in 1768, for John Finnie, tutor to the landowner, Matthew Orr.

Fintry (Stirling; Aberdeenshire) 'White house'. *Fionn* (Gaelic) 'white'; *tref* (Bittonic–Pictish) 'house', 'homestead'. Noted pre-1225 as Fyntryf. Fintray has the same derivation.

Fionnphort (Argyll & Bute) 'Fair harbour'. *Fionn* (Gaelic) 'fair', 'bright', *phort* 'harbour', 'beaching place'. This is one of the few unaltered Gaelic placenames in regular use.

Firth 'Sea inlet' or 'wide estuary', from *fjordr* (Old Norse). It was never borrowed into Gaelic and many fiord-names have been concealed by later gaelicisation with 'loch' added, see Inchard; in Scots form, usually metathesised to 'frith' until the late 19th century, it survives in the ten great Firths of the mainland coast as well in an Orkney placename of the same derivation.

Fitful Head (Shetland) 'Cape of the web-footed birds' (i.e. seabirds). *Fit* (Old Norse) 'foot', *fugl* 'bird'. 'Head' is an English translation of Old Norse *hofud*, 'headland'. An older suggested etymology is Old Norse *hvit-fjell* 'white hill'.

Fiunary (Highland) 'Fair shieling, or hill-pasture'. *Fionn* (Gaelic) 'bright', 'fair', *airidh* 'shieling', 'hill pasture'.

Fladda (Western Isles) 'Flat island'. The numerous Fladdas, Flodda and Fladday, are from *flat-ey* (Old Norse) 'flat island'.

Flannan Isles (Western Isles) 'St Flannan's isles'; in Gaelic, na h-Eileanan Flannach, from the 7th-century St Flannan.

Fleet, River (Dumfries & Galloway, Highland) 'Estuary'. *Fleot* (Old English) 'estuary'. The Fleet river and loch in Sutherland are 'flooding stream': *fljotr* (Old Norse) 'fleet', 'flood'.

Flemington (South Lanarkshire) 'Town of the Flemings'. Flemish immigrants entered Scotland from the twelfth century to the seventeenth. The several Flemingtons may denote places where Flemings (often weavers) settled, or in some cases may be back-named from the Scots surname Fleming, assumed by the incomers.

Flodigarry (Highland) 'Fleet garth'. *Flotr* (Old Norse) 'fleet' - expressive of water rather than ships; *gearraidh* (Gaelic) 'land between machair and moor'.

Flotta (Orkney) 'Fleet island', from *flotr* (Old Norse) 'fleet', *-ey* 'island'.

Fochabers (Moray) Perhaps 'lake-marsh', from *fothach* (Brittonic–Pictish) 'lake', and *aber* 'confluence', with the sense of 'marsh'. The terminal *-s*, found in 1514, Fochabris, appeared after 1325, when Fochabre is noted.

Footdee (Aberdeen; Fife). 'Peaty place'. *Fòid* (Gaelic) 'peat', *ait* (Gaelic locative suffix). Noted as Foty, 1337, and as Futismire, 1583. Loch Fitty in Fife is in a peaty area, and nearby Foodie Hill has the same derivation.

Fordyce (Aberdeenshire) 'South-facing slope'. *Faithir* (Gaelic) 'shelved or terraced slope', and *deas* 'south'.

Forfar (Angus) Possibly 'watching hill'. *Faithir* (Gaelic) 'terraced slope', *faire* 'watchfulness', 'sentinel'. The nearby Hill of Finhaven would have been a suitable place for such a lookout. It was recorded as Forfare c. 1200.

Forres (Moray) 'Below the bushes'. *Far* (Gaelic) 'below', and *ras* 'shrubs', 'underwood'. Recorded as Fores 1187; Forais 1283. The Gaelic name is Farrais. A connection with the local tribe of the Boresti, mentioned in CE 84 by Tacitus, has also been suggested.

Forsinard (Highland) 'Waterfall on the height'. *Fors* (Old Norse) 'waterfall'; *an* (Gaelic) 'of the', *àird* 'height'. It is higher than Forsinain (Gaelic *an fháin* 'low').

Fort Augustus (Highland) Formerly known as Kilchomain (Scottish Gaelic *cille Chumainn*, 'St Colman's church'), it was renamed after 1715 in commemoration of William Augustus, duke of Cumberland.

Fort George (Highland) Fort and village named after King George II, 1746.

Fort William (Highland) Originally the settlement of Inverlochy (*see* Lochy). A fort was established here in 1655, called Gordonsburgh, after the duke of Gordon on whose land it had been built. Shortly afterwards, its name changed briefly to Maryburgh, after Queen Mary II, co-sovereign with King William III. Finally, in 1690 it was renamed Fort William, after the king.

Forteviot (Perth & Kinross) The first element may be *faithir* (Old Gaelic) 'terraced or shelved slope'; but has also been linked to the Verturiones tribe. The name is not fully explored. In the *Pictish Chronicle*, c. 970, it is recorded as Fothuir-tabaicht, and is not related to the Border Teviot.

Forth, River and **Firth** (Stirling; North Lanarkshire) Of uncertain origin. Tacitus's first-century account calls the Forth Bodotria. A Brittonic or pre-Brittonic name **Voritia* has been conjectured, of obscure meaning. 'Forth' is recorded from the 12th century. It has not been preserved in Gaelic, where the river-name is Abhainn Dubh, 'black water'. The Vikings also knew it as Myrkvifjord, 'dark firth'. The North Lanarkshire village of Forth may have a different origin, perhaps from a Brittonic word related to Welsh *fford*, 'road'.

Fortingall (Perth & Kinross) Perhaps 'Fortified church'. **Fartair* (Old Gaelic) 'fortress', or perhaps *faithir* (see Forteviot) with *cill* (Gaelic) 'church'. Noted as Forterkil, c. 1240. In Gaelic it is Fartairchill. It has yet to be fully explained.

Fortrose (Highland) Perhaps 'Beneath the headland'. *Fo* (Gaelic) 'beneath', *ros* 'headland.' The first part may be cognate with Forres; the second part is also present in the neighbour-village of Rosemarkie. The form Forterose is found in 1455. The latter-day Gaelic name is A' Chananaich, 'town of the canons'.

Foss (Perth & Kinross) 'Stance', 'cattle station'. *Fas* (Gaelic) 'stance'.

Foula (Shetland) 'Bird Island'. *Fugl* (Old Norse) 'bird', *ey* 'island'.

Foulis (Highland; Perth & Kinross) 'Small stream'. *Foghlais* (Gaelic) 'lesser stream'. The same derivation applies to Easter and Wester Fowlis, in Perthshire, noted as Foulis, 1147. The name is found in other formations, as in Powfowlis, 'stream pools', near Stirling.

Foyers (Highland) 'The Brink'. *Fóir* (Gaelic) 'edge', 'brink'.

Fraserburgh (Aberdeenshire) 'Fraser's town'. Originally called Faithlie, of uncertain derivation, it was renamed in 1592 after Sir Alexander Fraser, developer of the town. The second element is derived from *burh* (Old English) becoming Scots *burgh* 'town'. It is known to locals as 'the Broch'.

Freuchie (Fife) 'Heathery place'. *Fraochach* (Gaelic) 'heathery'; noted as Fruchy, 1508. Loch Freuchie, in Perth & Kinross, is 'heathery loch'.

Friockheim (Angus) Originally known as Friock Feus ('rents'), apparently after a Forfar bailie named Freke; *heim* (German) 'home', 'village', was added in 1830 by the landowner, John Anderson.

Fruin, River and **Glen** (Argyll & Bute) Possibly 'raging (stream)'. *Freoine* (Gaelic) 'rage'. Recorded as Glenfrone in the thirteenth century.

Furnace (Argyll & Bute) Kilns were set up here on Loch Fyneside for iron-smelting in the 18th century, using local wood fuel, hence the name. In Gaelic it is An Fhúirneis.

Fyne, River, Glen and **Loch** (Argyll & Bute) 'Stream of wine, or virtue'. *Fine* (Gaelic) 'wine'. The name, probably first given to the river, is found in its present form from 1555. The reference may be to an ancient holy site, but the complimentary description of water as wine is not uncommon in Gaelic usage.

Fyvie (Aberdeenshire) Possibly a 'path'. *Fiamh* (Gaelic) 'path'. Noted pre-1300 as Fyvyn.

G

Gairloch (Highland) 'Short loch'. *Gearr* (Gaelic) 'short', *loch* 'lake', 'loch'. The village at its head has usurped the name, rather than Kingairloch as might be expected, and the loch is often called Loch Gairloch. Gareloch on the Firth of Clyde has the same derivation.

Gairsay (Orkney) 'Garek's isle', from *Garek* (Old Norse personal name) and *ey* 'island'.

Galashiels (Borders) 'Shielings by the Gala Water'. The latter part of the name comes from *skali-s* (Old Norse) 'sheilings'. The source of Gala has been suggested as *galga* (Old English) 'gallows', but the origin may be Brittonic *gal gwy*, 'clear stream'. The name was recorded as Galuschel 1237, Gallowschel 1416.

Galloway (Dumfries & Galloway) 'Land of the stranger Gaels'. *Gall* (Gaelic) 'stranger' and *Ghaidhil* 'Gaels'. Noted in the *Pictish Chronicle*, c. 970, as Galweya. This tribal name was given by the Scots to the extreme south-west, settled by people of mixed Irish and Norse origins, who were regarded as 'foreigners'.

Galston (East Ayrshire) 'Village of the strangers'. *Gall* (Gaelic) 'stranger'; *tūn* (Old English) 'village'. Recorded as Gauston, 1260.

Gardenstown (Aberdeenshire) 'Garden's town'. A fishing port set up in 1720 by Alexander Garden of Troup, in the parish of Gamrie (perhaps Gaelic *camag* 'bay' and *righe* 'slope'.

Garelochhead (Argyll & Bute) 'At the head of the short loch'. *Gearr* (Scottish Gaelic) 'short'; *loch* (Scottish Gaelic) 'lake, loch'. The English suffix suggests a modern name, but Keangerloch, with the Gaelic prefix *ceann*, 'head', is found in the early 14th century.

Gargunnock (Stirling) Possibly 'rounded hill'. *Garradh* (Gaelic) 'enclosure'; *cnuic* or *duin-ock* (Gaelic) 'rounded hill', 'mound'. The village lies at the base of the Gargunnock Hills.

Garmouth (Moray) 'Short plain'. *Gearr* (Gaelic) 'short', here in the sense of 'narrow'; *magh* 'plain'.

Garnkirk (North Lanarkshire) A 12th-century form Leyngartheyn suggests the name has been rendered from *lann* (Brittonic) 'church', into Scots *kirk*, and *gartan* (Gaelic) 'little field'. In 1515 it was noted as Gartynkirk.

Garnock, River (Renfrewshire; North Ayrshire) 'Little noisy stream', from Gaelic *gair*, 'cry', with the diminutive suffix *-ag*.

Garry, River, Loch and **Glen** (Highland; Perth & Kinross) 'Rough river'. The name comes from a Celtic root **garu*, as does *garbh* (Gaelic) 'rough'. The Border river-name Yarrow has the same derivation.

Garscadden (Glasgow) 'Herring yard'. *Gart* (Gaelic) 'enclosure', *sgadain* 'herring'. Noted as Gartscadane in 1373.

Garscube (Glasgow) 'Corn-yard or field'. *Gart* (Gaelic) 'enclosure, field', and *sguab* 'sheaf'. Noted as Gartskube in 1457.

Gartcosh (Glasgow) 'Field with the hollow, or cave'. *Gart* (Gaelic) 'enclosure, field', and *còs* 'hollow, cavern'. Found as Gartgois, 1520.

Garth (Perth & Kinross) 'Corn field'. *Gart* (Gaelic) 'enclosure', 'field', can have the specific sense of 'standing corn', 'cornfield'. Gart-, Garty- are common prefixes for local and field names, with the sense of 'field', especially 'cornfield'. But see Gartocharn.

Gartnavel (Glasgow) 'Apple field'. *Gart* Gaelic) 'enclosure', 'field', *n* 'of the', *abhal* 'apples'.

Gartness (Stirling) 'Field (probably cornfield) by the water, or stream'. *Gart* (Gaelic) 'standing corn', 'cornfield', *nan* 'of the', *eas* 'water'.

Gartney, Strath (Stirling) 'Gartan's strath'. *Gartán* (Gaelic personal name, perhaps related to Pictish Gartnait), *srath* (Gaelic) 'wide valley'. 15th century forms like Strongarnay suggest the original first element was Gaelic *sron* 'nose', 'projection'.

Gartocharn (West Dunbartonshire) 'Place of the humped hill'. *Garradh* (Gaelic) 'enclosure', 'place'; *chairn* (Gaelic *càrn*) 'humped hill'. Recorded as Gartcarne in 1485.

Gartsherrie (North Lanarkshire) 'Colt field'. *Gart* (Gaelic) 'enclosure', 'field', *searraigh* 'colts'. Recorded as Gartsharie, 1593.

Garve (Highland) 'Rough ground'. *Garbh* (Gaelic) 'rough'.

Garvelloch Isles (Argyll & Bute) 'Rough or rocky isles'. *Garbh* (Gaelic) 'rough', and *eileach* 'rock'. Noted 1390 as Garbealeach.

Gask (Aberdeenshire; Highland; Perth & Kinross) 'Tongue or tail of land'. *Gasg* (Gaelic) 'tail', and *gasgan* 'plateau tailing to a point'. Gask is also found as an element in numerous placenames.

Gatehouse of Fleet (Dumfries & Galloway) 'Roadhouse on the (Water of) Fleet'. *Geata-hus* (Old English) 'roadhouse', and see Fleet. The town was founded in 1790, but such travellers' hospices were monastic in origin and thus pre-Reformation.

Gateside (Angus; other areas) 'Roadside place or area', from Scots *gait* 'road'. Scots 'side' can have the sense of 'area' or 'district' as well as the current meaning. Numerous farms and hamlets have this name, mostly south of the Highland line.

Georgemas (Highland) There was an annual cattle market here on St George's Day, April 23, and the name simply means 'St George's mass, or feast'.

Giffnock (East Renfrewshire) 'Little ridge'. *Cefn* (Brittonic) 'ridge', *oc* (diminutive suffix) 'little'.

Gifford (East Lothian) The name appears to come from the Norman family name Gyffard, noted 1580 as Giffordiensis, 'of the Giffards'.

Gigha (Argyll & Bute) Possibly 'God's Isle'. *Gud* (Old Norse) 'God', *ey* 'island'. A derivation from Old Norse *gjá* 'gap', 'rift', has also been proposed. Recorded as Gudey 1263, Geday 1343, Gya 1400, Giga 1516.

Gilmerton (Edinburgh; other areas) 'Gilmour's farm'. *Gille Moire* (Gaelic personal name 'servant of Mary'); with *tūn* (Old English) 'farm'. Shown as Gillmuristona c. 1200, this name reveals a Gaelic revival in the Lothian area, with Gaelic personal names attached to what had been Anglian settlements. Numerous farms share the name, not always with such a long history.

Girvan (South Ayrshire) This much debated name has been related to Brittonic *gerw* 'rough', and *afon* 'river'. 'Bright garden' has also been proposed, from Gaelic *gàradh* 'garden', *fionn* 'fair', with an anachronistic resemblance to the Ptolemaic name Vindo-gara, CE 150. It was recorded as Girven in 1275.

Glamis (Angus) 'Wide gap'. *Glamhus* (Gaelic) 'wide gap', 'vale'. Records show Glammes 1187, Glammis 1251.

Glasgow 'Place of the green hollow'. *Glas* (Brittonic) 'green', and *cau* 'hollow'. The Gaelic form is Glaschu. Some authorities considered that a 'familiarative' sense was implied, hence the popular rendering 'dear green place'. Recorded as Glasgu in 1116.

Glass, River, Glen, Loch (Highland; Dumfries & Galloway) The noun *Glas* in Gaelic has several meanings, including 'water'; as an adjective *glas* means 'grey', 'green', 'pale'. There are numerous Glass rivers, and it is an element in many other names; *see* Douglas, Kinglass, etc.

Glenalmond (Perth & Kinross) 'Glen of the river'. *Gleann* (Gaelic) 'glen', 'valley'; and see Almond.

Glencoe (Highland) Probably 'the narrow glen'. *Gleann* (Gaelic) 'glen', 'valley', *cumhann* 'narrow'. Noted in 1343 as Glenchomure, then as Glencole, 1494, probably based on Gaelic *caol* 'narrow'. The modern Gaelic name is Gleann Comhann.

Glendaruel (Argyll & Bute) There are differing explanations and perhaps the original form has become too distorted to be recognisable. The elements *Gleann* (Gaelic) 'glen'; and *dà*, 'two', are clear, but the latter part has been variously derived from Gaelic *ruadhail*, 'red spots', and *ruadha*, 'points', 'headlands'. In 1238 it was recorded as Glen Da Rua.

Gleneagles (Perth & Kinross) 'Glen of the church'. *Gleann* (Gaelic) 'glen', and *eaglais* 'church'. It was recorded as Gleninglese around 1165.

Glenelg (Highland) This has been taken for 'glen of Ireland'. *Gleann* (Scottish) 'glen'; *Ealg* Old Gaelic name for Ireland. Like Elgin, the reference here would be a commemoration of the motherland by early Gaelic-speaking settlers. It has also been derived from *eilg* (Old Gaelic) 'noble' (compare Wyvis). Recorded as Glenhelk in 1282.

Glenfinnan (Highland) Possibly 'glen of (St) Finan'. *Gleann* (Gaelic) 'glen'; *Finan* (Irish Gaelic personal name, from *fionn*, 'fair') a seventh-century monk from Iona.

Glenkens, The (Dumfries & Galloway) 'Glen of the waterhead' or 'white river'. *See* Ken. The plural ending has been ascribed to the four medieval parishes forming the district; in 1181 it was noted as Glenkan.

Glenlivet (Moray) Apparently 'glen of the slippery smooth place'. Gaelic *Gleann* 'glen', *liobh* 'slippery', 'smooth', *ait* 'place'.

Glenrothes (Fife) A modern name created when the town was established in 1948 as a mining centre. There is no glen; the second part acknowledges the earls of Rothes as local landowners, and their former Rothes Colliery.

Goat Fell (North Ayrshire) 'Goats' hill'. A hill name from Old Norse *Geitar* 'goats', and *fjall* 'hill'. On Blaeu's map it is shown as Keadefell Hil. The Gaelic name is Gaoda-bheinn.

Goil, Loch (Argyll & Bute) Perhaps 'loch of the Gall, or stranger'; or 'of the rock'. *Loch* (Gaelic) 'lake', 'loch', *goill* 'of the stranger' or 'of the rock'. Glen Gyle would seem to have the same possible origins.

Golspie (Highland) Apparently 'Gulli's farm'. *Gulli* (Old Norse personal name); *by* 'farmstead'. The suffix is rare on the mainland. Recorded as Goldespy 1330; Golspi 1448.

Gorbals (Glasgow) The origin remains uncertain. One authority has suggested a derivation with reference to *gorr balk-r* (Old Norse) 'built walls'. It is recorded in a document of 1521 as Gorbaldis.

Gordon (Borders; Aberdeenshire) 'Great fort'. *Gor* (Brittonic prefix to intensify meaning), *din* 'fort'. The name originated in the south and was transferred northwards by the family of Gordon in the 13th century. Gourdon is probably the same name.

Gorebridge (Midlothian) Possibly the 'bridge at the wedge-shaped land'. *Gora* (Middle English) 'triangular piece of land', with English 'bridge'. For a Scots form see Gushetfaulds.

Gorgie (Edinburgh) Probably 'big field'. *Gor* (Brittonic intensifying prefix) 'big', with *cyn* 'field'.

Gourock (Inverclyde) Suggested as the 'place of the hillocks'. *Guireag* (Gaelic, a form of *guirean*) 'pimples' in the sense of small hills.

Govan (Glasgow) 'Mound', 'hillock' from Brittonic *go* 'small', *ban* 'hill' referring to the one-time artificial mound used as a meeting-place. This is considered more likely than *cefn* (Brittonic) 'ridge' and other proposed derivations. Recorded as Ovania, 8th century, Guven 1147, Gvuan 1150, Gwuan 1518.

Gowrie (Perth & Kinross) Perhaps 'Gabràn's land'. *Gabràin* (Old Gaelic personal name) 'of Gabràn'. Gowrie with Atholl formed one of the seven major divisions of Pictland. The name is found as Gowrin, 1120. Carse of Gowrie is *carse* (Scots from Old Norse *kerss*) 'low-lying land by a river'.

Graemsay (Orkney) 'Grim's isle'. *Grim* (Old Norse personal name), and *ey* 'island'.

Grahamston (Falkirk) A relatively modern name using the well-established suffix *-ton*; indicating a township. Previously called Graham's Muir, after Sir John Graham, killed in the Wars of Independence, 1298.

Grampian Mountains (Highland; Moray; Perth & Kinross) Apparently a mis-rendering of Graupius (Mons Graupius, site of a battle between Romans and Caledonians in AD 84, recorded by Tacitus), first written by Hector Boece in his *History of Scotland*, 1526. The derivation of 'Graupius' is unknown. The historic name is The Mounth, from Gaelic *monadh* 'mountain'. See Monadhliath.

Grandtully (Perth & Kinross) 'Thicket on the hill'. *Cardden* (Pictish) 'thicket'; *tulach* (Gaelic) 'hill'. Noted prior to 1400 as Garintully.

Grange (Fife; other districts) This name indicates the presence of a pre-Reformation (mid-16th century) grange or barn, used for storing produce belonging to an abbey. Many parishes in Scotland were chartered to abbeys, and *-grange* occurs as an element in many names.

Grangemouth (Stirling) 'Mouth of the Grange Burn'. The town stands at the mouth of the Grange Burn, named after the nearby grange for Newbattle Abbey.

Grantown-on-Spey (Highland) Originally built in 1766 as a planned village for the local landowner Sir James Grant, after whom it is named.

Granton (Edinburgh) 'Green hill'. *Gren* (Old English) 'green', *dún* 'hill', 'rise'. Noted around 1200 as Grendun.

Greenan (Argyll & Bute) 'Sunny place'. Gaelic *Grianán* 'sunny place', from *grian*, 'sun'. A frequent local name for a sun-facing slope. Grenan in Bute has the same source, as do Grennan and Bargrennan in Dumfries & Galloway.

Greenlaw (Borders) 'Green hill'. *Gren* (Old English) 'green', and *hláew* 'hill', giving Scots 'law'. Noted as Grenlawe, 1250.

Greenock (Inverclyde) The traditional origin from Gaelic *Grianàig* 'at the sunny knoll' has been challenged by a Brittonic one as 'gravelly place', cognate with Welsh *graenoc*. Noted in 1400 as Grenok. The name is also found as a local one near Callander and Muirkirk.

Gretna (Dumfries & Galloway) Possibly place of the 'gravelly haugh'. *Greoten* (Old English) 'gravelly', *halth* 'haugh' or 'fertile land enclosed by a river bend'. Records show Gretenho 1223; Gretenhowe 1376; Gretnay 1576.

Grimshader (Western Isles) 'Grim's farm'. *Grimr* (Old Norse proper name); *seadair* (Gaelic from Old Norse *saetr*) 'farm'.

Gruinard (Highland) This has been taken as 'Split firth', *grein* (Old Norse) 'split', 'divided', and *fjordr* 'firth'. But, like Gruinart in Islay, it may be 'shallow firth', from *grunna* (Old Norse) 'shallow', which also fits the location.

Gullane (East Lothian) 'Shoulder hill'. Perhaps from *guallan* (Gaelic) 'shoulder', with the local hill, Gullane Law, in mind. The emphasis on the first syllable makes a *-linn*, 'lake' ending unlikely. Noted as Galin, c. 1200.

Gushetfaulds (Glasgow) 'Cattle or sheep-folds in the gore'. *Gushet* (Scots) 'triangular corner', 'gore', *fauld* 'fold'.

Guthrie (Angus) 'Windy slope', from Gaelic *gaoth* 'wind', *righe* 'slope', 'summer sheiling'. Noted in 1359 as Gutherie.

Gyle (Edinburgh) Perhaps 'place of the Gall, or strangers', from Gaelic *goill* 'of the stranger'. For another possibility see Goil, Loch.

H

Haddington (East Lothian) 'Hada's people's place'. *Hada* (Old English personal name); *inga* 'people's', *tūn* 'settlement', 'farm'. Recorded as Hadynton 1098; Hadintun and Hadingtoun 1150.

Hailes (Edinburgh) 'Halls'. Likely to derive from *heal* (Old English) 'hall'. Recorded as Hala, c. 1150. North and South Halls are recorded in the 14th century.

Halkirk (Highland) 'High church'. *Há* (Old Norse) 'high', *kirkju* 'church'. Noted in 1222 as Hakirk, metathesised by 1621 to Halkrig; the process has also affected the Gaelic form Hacraig.

Halladale, River and **Strath** (Highland) Either 'holy dale' from *helg* (Old Norse) 'hallowed', and *dalr* 'valley'; or 'Helgi's dale' from an Old Norse personal name. Recorded in 1222 as Helgadall.

Hamilton (South Lanarkshire) From the Norman-French name de Hameldon, renamed from Cadzow by the first Lord Hamilton in the 15th century.

Hamnavoe (Shetland) 'Harbour of the bay'. *Hamn* (Old Norse) 'harbour', and *vágr* 'bay'. Recorded in Old Norse writings as Hafnarvag. *See* Stromness.

Harris (Western Isles) 'Higher island'. *Haerri* (Old Norse) 'higher', from *hár*, seems the likeliest derivation, though the -s ending is not explained. The Gaelic rendering is Na h-Earra, with the definite article. Recorded as Heradh 1500; The Harrey 1542; Harreis 1588.

Hatton (Aberdeenshire; other districts) 'Hall farm'. *Hall* (Old English) 'manor house'; *tūn* 'farmstead', giving Scots 'toun'. As with many Scots words ending in *-all*, the *-l* sound has been lost, in this case in spelling as well as pronunciation.

Hawick (Borders) 'Hedge-enclosed settlement'. *Haga* (Old English) 'hedge', *wic* 'settlement', 'farm'. Recorded as Hawic in the twelfth century.

Hebrides (Western Isles) Noted by the Roman Pliny in CE 77 as *Haebudes*, and referring to the Inner Hebrides, it has also been applied to the 'Long Island' or Outer Hebrides. The present name is also the result of a misreading of *ri* for *u* in the transcription of old manuscripts, at some point before the 16th century. The derivation is unknown. The Gaelic name is Innse Gall, 'isles of the strangers', given when they were under Norse occupation. In Old Norse they were Sudreyar, 'southern islands'.

Helensburgh (Argyll & Bute) 'Helen's town'. Named for the wife of Sir James Colquhoun of Luss, who in 1752 bought land here.

Hell's Glen (Argyll & Bute) English for *Ifrinn* (Gaelic) 'Hell'. But it is likely that the true meaning is the opposite, from Gaelic *aifrionn* 'chapel', 'place of offerings'. *See* Inchaffray.

Helmsdale (Highland) 'Hjalmund's dale'. *Hjalmund* (Old Norse personal name), and *dalr* 'valley', 'dale'. Recorded in the *Orkneyinga Saga* of c. 1225 as Hjalmunddal; in 1290 noted as Holmesdale.

Heriot (Borders) Perhaps 'strategic pass'. *Here* (Old English) 'army', *geat* 'hill-pass', indicating a pass through which an army could march abreast. An alternative is *here-geatu*, 'army equipment', perhaps a form of service required from the tenant. Noted around 1200 as Heryt.

Hermiston (Edinburgh) 'Herd's place'. *Hirdmannis* (Old English) 'Herdsman's', *tūn* 'settlement', 'place'. Recorded as Hirdmannistoun in 1233.

Highlandman (Perth & Kinross) The road here was known as Highlandman's Loan, since it was the route south for Highland cattle drovers.

Hilton (Highland; other regions). 'Village on or by the hill'. In Gaelic *Bail' a' Chnuic* 'village by the hill'. Sometimes given an additional locative designation, as with Hilton of Cadboll in Easter Ross.

Hirta (Western Isles) As Hirt or Hiort, the Gaelic name of the St Kilda islands; in English it has become the name of the main island. Its derivation may be Old Norse *hirtir*, 'stags', recorded from 1202, presumably a reference to the horn-like peaks of the islands; or as Old Norse *hjorthey* 'herd island'; or as Old Irish *hirt* 'death', with the sense that this remote archipelago was the gateway to the underworld. *See* St Kilda.

Holy Loch (Argyll & Bute) English for Gaelic *An Loch Seunnta*. Apparently so called from its association with St Mund, follower of St Columba. See Kilmun.

Holywood (Dumfries & Galloway) Originally Darcongall: 'wood of St Comgall' (Gaelic *doire* 'copse') and 6th century personal name; rendered into English as 'holy wood'.

Hope, Ben and **Loch** (Highland) 'Bay'. *Hop* (Old Norse) 'bay', 'shelter place'. The mountain takes its name from the loch, as does the township of Hope.

Hopeman (Moray) The village, founded 1805, is on a site previously known for unrecorded reason as 'Haudment', French *haut* 'high', and *mont* 'hill'.

Hourn, Loch (Highland) 'Furnace' or 'gully'. *Sòrn* (Gaelic) 'snout', 'furnace,' 'concavity'. The Gaelic name is Loch Shuirn.

Houston (Renfrewshire) 'Hugo's farmstead'. *Hugo* (Old English personal name), *tūn* 'farmstead'. In the 12th century it was the property of Hugo de Paduman, noted around 1200 in Latin as Villa Hugonis.

Howe o' the Mearns (Aberdeenshire) This district, between Laurencekirk and Stonehaven, is literally the 'hollow of the Mearns', from Scots *howe* 'hollow', a direct translation of Gaelic *Lag na Maoirne*. See Mearns.

Howgate (Midlothian) Perhaps 'road in the howe'. *Howe* (Scots) 'hollow', *gait* 'road'.

Hoy (Orkney) 'High island'. *Hár* (Old Norse) 'high', and *ey* 'island'. It was recorded as Haey in the *Orkneyinga Saga*, c. 1225, Hoye, 1492.

Humbie (Fife; East Lothian) 'Dog's town'. *Hund* (Old English) 'dog' probably used as a nickname, and *by* (Old Norse) 'settlement: place'. There are at least five Humbies in Southern Scotland.

Huntly (Aberdeenshire, Perthshire) 'Huntsman's wood'. *Hunta* (Old English) 'huntsman', *leah* 'wood'. Noted as Huntlie in 1482. Originally a Borders placename (there is a Huntlywood near Earlston), it was transferred north in the 13th century by the Gordon family.

Hutchesontown (Glasgow) The land here was purchased by the Hutcheson brothers, founders of Hutcheson's Hospital (1639) and developed in the 1790s.

Hyndland (Glasgow) 'Back land'. A direct Scots translation of the Gaelic form *cul tìr*, 'back land'. Noted as 'the hynde laude', 1538. See Culter.

I

Ibrox (Glasgow) 'Ford of the badger'. *Ath* (Gaelic) 'ford', and *bruic* 'badger's'.

Inchaffray (Perth & Kinross) 'Isle, or water-meadow, of the chapel'. *Innis* (Gaelic) 'island', 'water-meadow', *aifrionn* 'chapel', 'place of offering'. Noted c. 1190 as Incheaffren.

Inch Cailleach (Argyll & Bute) 'Isle of the old woman'. *Innis* (Gaelic) 'island', and *cailleach* 'old woman'. Recorded as Innischallach, 1411. One of the islands in Loch Lomond.

Inchcape (Highland; Angus) 'Isle, or water-meadow, of the block or head'. *Innis* (Gaelic) 'island', and *ceap* 'block', 'head'. Old Norse *skeppa* 'basket', has also been suggested.

Inchcolm (Fife) 'Island of (St) Columba'. *Innis* (Gaelic) 'island; *Columba* (Latin 'dove' giving Gaelic *Colum*).

Inchinnan (Renfrewshire) 'Isle of St Finnan'. *Innis* (Gaelic) 'island'; *Finnén* (Old Gaelic personal name). Noted as Inchenan in 1158. See Kilwinning.

Inchkeith (Midlothian) Gaelic *innis*, 'island'; For the second element Bede's *Ecclesiastical History of the English People*, of 731, refers to 'Giudi', which may be a personal or tribal name. Noted as Ynchkeyth, 1461.

Inchmahome (Stirling) Long identified as 'Isle of St Colman', now Columba is considered more likely. *Innis* (Gaelic) 'island', *mo* 'my'; *Colmóc* (Old Irish personal name). Noted in 1238 as Inchmaquhomok (see Portmahomack).

Inchnadamph (Highland) 'Isle, or water meadow of the oxen'. *Innis* (Gaelic) 'water meadow', *na* 'of the', *daimh* 'stag' or 'oxen'. 'Isle' here describes an area of arable limestone soil in the midst of older rocks.

Inchture (Perth & Kinross) Possibly 'hunting meads'. *Innis* (Gaelic) 'water-meadow', *a* 'of the', *thòire* 'pursuit', 'chase'.

Ingliston (Edinburgh) 'Ingialdr's farm'. *Ingialdr* (Old Norse personal name); *tūn* (Old English) 'farm', though Scots *Inglis*, 'of the English(man)' is equally possible.

Inkster (Orkney) 'Ing's farm'. *Inga* (Old Norse proper name), *saetr* 'farmstead'. Ingsetter has exactly the same derivation.

Innellan (Argyll & Bute) 'Place of islands'. *An-eilean* (Gaelic) 'island place'. Noted 1571 as Inellane.

Innerleithen (Borders) 'Confluence of the River Leithen'. *Inbhir* (Gaelic), Leithen a river-name related to *leathann*, 'broad', with the sense of broad surrounding slopes. Recorded as Innerlethan around 1160.

Insch, also **Insh** (Aberdeenshire; Highland) 'Meadow'. *Innis* (Gaelic) 'water-meadow'.

Inver- This very common prefix indicates a place where two rivers converge, or where a river enters a loch or the sea. The Gaelic form *Inbhir* is derived from an early Celtic root **eni-beron*, 'in-bring'. *See also* Aber-.

Inver (Highland; Perth & Kinross) *Inbhir* (Gaelic) 'river-mouth', 'confluence'. Here the name is used on its own (though the former was once Inverlochslin).

Inveramsay (Aberdeenshire) Perhaps 'Confluence of the dirty stream'. Gaelic *Inbhir* 'confluence'; *al* from a pre- or early-Celtic root **alauna.* 'flowing'; *mosaich* 'dirty'. Noted as Inveralmeslei, 1260.

Inveraray (Argyll & Bute) 'Mouth of the River Aray'. *Inbhir* (Gaelic) 'river-mouth'; *Aray* (pre-Celtic river-name) probably means 'smooth-running', found throughout Europe in many variant forms: Aar, Ahr, Aire, Ara, Ayr, Oare, Ore, etc.

Inverbervie (Aberdeenshire) 'Mouth of the Bervie Water'. *Inbhir* (Gaelic) 'river-mouth'; Bervie (probably Brittonic river-name similar to Welsh *berw*) 'boiling', 'seething'. The original name was Aberbervie (Haberberui, 1294), later gaelicised.

Inverewe (Highland) 'Mouth of the River Ewe'. See Kinlochewe.

Invergarry (Highland) 'Mouth of the River Garry'. *Inbhir* (Gaelic) 'river-mouth', Garry (Gaelic river-name derived from *garbh* 'rough').

Invergordon (Highland) A fabricated name given around 1760 after the town's founder, Sir Alexander Gordon. Previously named Inverbreckie, from the local Breckie (from Gaelic *breac*, 'speckled') Burn.

Invergowrie (Perth & Kinross) There is no Gowrie river; the name, though much older, seems to be formed on the same basis as Invergordon; 'Gowrie on the Tay estuary'. Recorded in 1124 as Invergourin. *See* Gowrie.

Inverkeilor (Angus) 'Mouth of the clay stream'. *Inbhir* (Gaelic) 'river-mouth', *cil* 'red clay', *dobhar* 'stream'. Recorded as Innerkeledur, c. 1200. *See* Rankeilour.

Inverkeithing (Fife) 'Mouth of the Keithing Burn'. *Inbhir* (Gaelic) 'river-mouth'; *Keithing* (Brittonic river-name derived from *coed*, 'wood'). Early records show Hinhirkethy c. 1050; Innerkethyin 1114.

Inverkip (Inverclyde) 'River mouth by the crag'. *Inbhir* (Gaelic) 'river-mouth', and *ceap* 'block', 'head'. Noted as Innyrkyp, c. 1170.

Inverkirkaig (Highland) 'Mouth of the kirk-bay stream.' *Inbhir* (Gaelic) 'river-mouth' added to *kirkja* (Old Norse) 'church', and *vik* 'bay'. In this case the settlement has untypically given its name to the river.

Inverness (Highland) 'Mouth of the River Ness'. *Inbhir* (Gaelic) 'river-mouth'; *Nis* (pre-Celtic river-name of undetermined origin). Recorded as Inuernis in 1171. See Ness.

Inversnaid (Stirling) 'Mouth of the needle-stream'. The Snaid burn, flowing into Loch Lomond, appears to be connected with *snàthad* (Gaelic) 'needle'.

Inverurie (Aberdeenshire) 'Confluence of the River Urie'. *Inbhir* (Gaelic) 'confluence'; Urie (Gaelic river-name), *see* Urie. Noted in 1175 as Enneroury. Here the Urie flows into the River Don.

Iona (Argyll & Bute) The original form was Ioua, but manuscript writers supplanted the *u* with *n*. The most probable derivation is from Old Irish *eo* 'yew', a tree always associated with holy places. When the area passed into Norse control, the name appears to have undergone some confusion with Norse *-ey* 'island'. Around 1100 the name is noted as Hiona-Columcille, and until around 1800 the island was known as Icolmkill: *ey* (Old Norse) 'isle'; *Columcille* (Irish Gaelic personal name) 'dove of the church'.

Irongray (Dumfries & Galloway) 'Land portion of the horse stud'. *Earran* (Gaelic) 'portion'; *na* 'of', *greigh* 'horse herd', 'stud'. The oldest known record, Drungray, 1298, is taken as a scribe's mistake; it is Yrnegray when found in 1468.

Irvine (North Ayrshire) Possibly place of 'the white (river)'. *Yr* (Brittonic) 'the', *(g)wyn* 'white'. Early records show Yrewyn c. 1140, Irvin 1230.

Isla, River and **Strath** (Perth & Kinross; Moray) A river-name tentatively traced back to a pre-Celtic root form **il* or **eil*, with the meaning of 'rapid-moving'. The southerly Isla is noted in 1187 as Strathylaf.

Islay (Argyll & Bute) Possibly 'Ile's island'. *Ile* (Old Norse personal name), *ey* 'island'. The interpolation of the *s*, modelled on English 'isle', is relatively recent. Recorded as Ilea c. 690, Ile 800.

J

Jedburgh (Borders) 'Town by the Jed Water'. The first element is that of the River Jed, probably derived from a form of *gweden* (Brittonic) 'winding', 'twisting'. The second is *burh* (Old English) 'town'. An earlier name was Jedworth, signifying 'enclosure by the Jed water' and still found in Bonjedward, a three-language hybrid of *bonn* (Gaelic) 'foot', *Jed*, and *worth* (Old English) 'enclosure', referring to the lower meadowland. Records confirm Gedwearde c. 800, Jaddeuurd c. 1140, Jeddeburgh 1160.

John o' Groats (Highland) Named after Jan de Groot, a Dutchman who came to live in Caithness in the late 15th century under the patronage of King James IV. The final -*s* is a reminder that the original form was 'John o' Groat's House'.

Johnstone (Renfrewshire; Dumfries & Galloway) The Renfrewshire town was founded by John Houston in 1782. The Dumfries-shire parish is 'John's settlement'. Scots *John* (personal name), *ton* 'farm', 'settlement' (from Old English *tūn*).

Joppa (Edinburgh) Area name from the 1780s, originally of a farm called after the Biblical Joppa (now Jaffa).

Jordanhill (Edinburgh; Glasgow) For these 16th- and 17th-century names, and for Edinburgh's Jordan Burn, a landowner's religious feeling seems the most likely explanation.

Juniper Green (Edinburgh) A 19th-century name, formerly Curriemuirend, the next place to the west being the village of Currie. First recorded in 1812.

Jura (Argyll & Bute) Apparently 'Doirad's island'. *Doirad* (Gaelic personal name), *ey* (Old Norse) 'island'. The Old Norse element replaced earlier Gaelic *eilean*, recorded as Doirad Eilinn in a document of 678.

K

Katrine, Loch (Stirling) Although the river flowing out of
Loch Katrine is the Achray Water, the name may
nevertheless be a river-name, from *cet* (Brittonic) 'wood',
and the same archaic root as may be found in 'Earn'.
Recorded as Ketyerne, 1463.

Keiss (Highland) 'Jutting place'. *Keisa* (Old Norse)
'protrude'.

Keith (Moray) The etymology is unclear; possibly the
Pictish personal name Cait, but more probably from *coit*
(Brittonic and Old Gaelic) 'wood', as in Dalkeith.
Recorded in 1203 as Ket. Fife Keith, beside the old
town, laid out in 1817, commemorates James Duff,
fourth earl of Fife. In earlier times the town was
sometimes referred to as Kethmalruf ('of St Maelrubha',
1220) and Ketmariscalli ('of the marischal', 1250), the
latter noting its association with the Keiths, hereditary
earls marischal of Scotland. The same family owned land
in East Lothian as Keith Hundeby (see Humbie).

Kells (Dumfries & Galloway) Perhaps 'wells' or 'springs',
from Old Norse *kell* 'spring'. The several Kells of Ireland
are from Irish Gaelic *Na Cealla* 'the (monastic) cells',
which is also possible here.

Kelso (Borders) Place of the 'chalk hill'. *Calc* (Old English)
'chalk, 'limestone', *hoh* 'hill'. There is still a part of the
town known as 'the chalkheugh'. Early records show
Calkou 1126; Kelcou 1158; Kelsowe 1420.

Kelty (Fife) 'Woods'. *Coilltean* (Gaelic) 'woods'. A record of
1250 shows Quilte.

Kelvinside (Glasgow) 'Narrow river'. Tradition takes the name from the River Kelvin, *caol abhainn* (Gaelic) 'narrow river'; a Brittonic origin is more likely; more analysis is needed. Scots *side* can mean 'area around'.

Kemnay (Aberdeenshire) 'Head of the plain'. *Ceann* (Gaelic) 'head', *a* 'of the', *maigh* 'plain'. The transposing of the *n* and *m* goes back at least to the 14th century, in the early years of which the region was forcibly 'scotticised' by Robert Bruce. Found as Camnay, 1348.

Ken, River and **Loch** (Dumfries & Galloway) Suggested as 'white (stream)', from *càin* (Gaelic) 'white', 'pure'. Alternatively Brittonic *pen* 'head (water)' gaelicised to Ken (*ceann* 'head'). See Glenkens.

Kenmore (Perth & Kinross) 'Great head'. *Ceann* (Gaelic) 'head', and *mòr* 'big'. Found in this form 1258.

Kennethmont (Aberdeenshire) 'St Alcmund's church'. Alcmund was an early bishop of Hexham, Northumberland, and this name is recorded from the 12th century as Kylalcmund, with *cill* (Gaelic) 'church'. Kynalcmund is also an old form. Later the name was misinterpreted and remodelled on the more familiar 'Kenneth' and *mont* 'hill'.

Kennoway (Fife) Probably 'place at the head', Gaelic *ceann* 'head', with two locational suffixes, '-ach' and '-in'. Kennachyn in 1160, Kenhacghy in 1276 .

Kentallen (Argyll & Bute) 'Head of the inlet'. *Ceann* (Gaelic) 'head', *an t'* 'of the', *saileinn* 'small inlet'.

Keppoch (Highland) 'Block,' or 'top'. *Ceap* (Gaelic) 'block, top'; with *-ach* (suffix indicating 'field').

Kerrera (Argyll & Bute) Perhaps 'copse island'. *Kjarbr* (Old Norse) 'copse', *ey* 'island'. Noted in 1461 as Carbery.

Kerry (Argyll & Bute; Highland) A name of different meanings. 'Fourth part' *Ceathraimh* (Gaelic) 'quarter', relating to early land divisions. Kerry in Wester Ross is from Old Norse *kjarr* 'copse', with *á* 'river'.

Kessock (Highland) 'St Kessoc's place'. Found as Kessok, 1437. Kessog was a 6th century Irish missionary more associated with the Luss district. In Gaelic it is Aiseag ('ferry') Cheiseig.

Kilbarchan (Renfrewshire) Probably 'the place of St Berchan's church'. *Cill* (Gaelic) 'church', *Berchan* (personal name) of a 7th-century Irish saint. Found in this form 1246.

Kilbirnie (North Ayrshire) Probably 'the place of St Brendan's church'. *Cill* (Gaelic) 'church'; *Brénaind* (Irish Gaelic personal name). Recorded as Kilbyrny in 1413.

Kilbowie (Renfrewshire) 'Yellow back'. *Cùl* (Gaelic) 'back' perhaps here implying the hill slope, and *buidhe* 'yellow'. 13th-century forms of the name show the *cul-* prefix, e.g. Cùlbuthe, 1233.

Kilbrandon (Argyll & Bute) 'Church of St Brandon'. *Cill* (Gaelic) 'church'; *Brénaind* (Irish Gaelic personal name).

Kilbrannan (Argyll & Bute) 'Strait of Brandon'. *Caol* (Gaelic) 'strait', 'kyle'; *Brénaind* (Irish Gaelic personal name). Noted as Culibrenin, 1549. Ignorance of the Gaelic meaning has prompted the later addition of 'Sound' to the name of this sea channel.

Kilbride (Argyll & Bute) 'St Bride's church'. Fifteen saints bore the name of Brìd, formerly that of a pagan goddess; and the name is found all across Scotland. *Cill* (Gaelic) 'church'; *Brìd* (Irish Gaelic personal name) 'Bride', 'Bridget'.

Kilconquhar (Fife) 'Church of Dúnchad. *Cill* (Gaelic) 'church', *Dúnchad* 8th century saint, initial 'd' modified to 'c' at an early stage. Recorded as Kilconcath, 1228, then Kinneuchar, 1699, which suggests *ceann* (Gaelic) 'head'; and *uachdair* 'of the upper ground', a false etymology. The Kil- form has prevailed, though not in pronunciation.

Kilcreggan (Argyll & Bute) 'Church on the little crag'. *Cill* (Gaelic) 'church', *creag* 'rock', 'crag', *an* (Gaelic diminutive suffix).

Kildary (Highland) 'Narrow wood'. Kil- here is *caol* (Gaelic) 'narrow', with *daire* 'oak wood'.

Kildonan (Highland; Argyll & Bute; Western Isles) 'Church of St Donnan'. *Cill* (Gaelic) 'church'; *Donnán* (Irish Gaelic personal name). St Donnan was murdered on Eigg in 617.

Kildrummy (Aberdeenshire) 'Head of the ridge'. *Ceann* (Gaelic) 'head', and *druim* 'ridge'. Recorded as Keldrumin, 1238, but known up to the 19th century as Kindrummie.

Kilkerran (Argyll & Bute) 'Church of St Ciaran'. *Cill* (Gaelic) 'church'; *Chiaráin* (Old Irish personal name). This is probably St Ciaran of Clonmacnoise in Ireland, who died in 549. Noted pre-1250 as Kilchiaran.

Killearn (Stirling) This name appears to have undergone a change similar to Kilconquhar, from an earlier form based on Gaelic *cinn*, 'at the head of', and *earrain*, 'of the land-portion'. The form Kynerine is found from c. 1250, but by c. 1430 it was Killerne, with Gaelic *cill*, 'church', replacing the *cinn* form.

Killearnan (Highland) 'Church of St Earnan'. *Cill* (Gaelic) 'church', *Iurnáin* (Gaelic form of Irish Gaelic personal name). Found as Kyllarnane, 1569.

Killiecrankie (Perth & Kinross) 'Wood of aspen trees'. *Coille* (Gaelic) 'wood', and *creitheannich* 'of aspens'.

Killin (Stirling) Probably 'place of the white church'. *Cill* (Gaelic) 'church', *fionn* 'white'. Noted as Kyllyn, 1318.

Kilmacolm (Inverclyde) Church of my Columba'. *Cill* (Gaelic) 'church', *mo* 'of my' *Coluim* (personal name) referring to the most famous early Irish–Scots saint. The addition of *mo* here denotes dedication. Kilmacolme is recorded in 1205.

Kilmaluag (Argyll & Bute) 'Church of Mo-Luoc'. *Cill* (Gaelic) 'church', *mo* 'of my'; *Lugaidh* (Old Irish personal name). This missionary saint's name is often given wrongly as Moluag; the *mo* element being separate.

Kilmany (Fife) 'Maine's Church'. *Cill* (Gaelic) 'church', *many* refers to a proper name, perhaps the 6th-century St Maine though he is not otherwise commemorated in Scotland. See Dalmeny.

Kilmarnock (East Ayrshire) 'Church of my little St Ernan'. *Cill* (Gaelic) 'church', *mo* 'of my', *Iarnan* (personal name) reputedly priest and uncle of St Columba, *-oc* (diminutive suffix). The name was recorded as Kelmernoke in 1299.

Kilmaronock (East Dunbartonshire) 'Church of my little St Ronan'. Apart from the different personal name, the derivation is the same as that of Kilmarnock. Noted as Kilmerannok, c. 1325.

Kilmartin (Argyll & Bute; Highland) 'Church of St Martin'. *Cill* (Gaelic) 'church', *Mhàrtuinn* (Gaelic form of 'Martin').

Kilmaurs (East Ayrshire) 'Church of St Maurice'. *Cill* (Gaelic) 'church', *Mauruis* (Gaelic form of Maurice). Noted in 1413 as Sancte Maure.

Kilmelfort (Argyll & Bute) 'Church of the sandy firth'. *Cill* (Gaelic) 'church'; *melr* (Old Norse) 'sand', *fjordr* 'firth', 'loch'.

Kilmuir (Highland) 'Mary's church'. *Cill* (Gaelic) 'church', *Mhuire* (Gaelic proper name) 'Mary's'. Noted as Kilmor, 1296.

Kilmun (Argyll & Bute) 'Church of St Mund'. *Cill* (Gaelic) 'church'; *Mundu* (Irish Gaelic personal name) a disciple of St Columba. Recorded as Kilmun 1240; Kilmond 1410. *See* Holy Loch.

Kilninver (Argyll & Bute) 'Church at the confluence'. *Cill* (Gaelic) 'church', *an* 'of the', *inbhir* 'confluence'. Noted 1250 as Kyllivinor.

Kilrenny (Fife) 'Church of the bracken'. *Cill* (Gaelic) 'church', *reithneach* 'bracken'. Found as Kilrinny, c. 1160.

Kilsyth (North Lanarkshire) Possibly 'church of St Syth'. *Cill* (Gaelic) 'church'; Syth (personal name) not recorded before the 16th century. An alternative has been suggested in *saighde* (Gaelic) 'arrows'. Recorded as Kelvesyth in 1210 and Kelnasythe in 1217, possibly suggesting a link with the River Kelvin whose source is nearby at Kelvinhead.

Kiltarlity (Highland) 'Church of Talorcan'. *Cill* (Gaelic) 'church', *Taraghlain* (Gaelic form of Pictish personal name) 'Talorcan's'. The name was recorded around 1225 as Kyltalargy.

Kiltearn (Highland) 'Church of the Lord (God)'. *Cill* (Gaelic) 'church', *Tighearna* 'Lord'. Noted as Keltierny, 1226.

Kilwinning (North Ayrshire) 'Church of St Finnian'. *Cill* (Gaelic) 'church'; *Finnian* (Irish Gaelic personal name). Recorded in 1202 as Kilvinnin.

Kinbrace (Highland) 'Seat of the chief (literally, 'of the brooch'). *Ceann* (Gaelic) 'head', 'head place', *na* 'of', *bhraiste* 'brooch', 'badge'.

Kincardine (Fife; Aberdeenshire; Highland; Tayside) 'At the head of the wood'. *Cinn* (Gaelic locative of *ceann*) 'at the head of'; *cardden* (Pictish) 'wood', 'thicket'. Recorded as Kynge Carden, 1295.

Kincardine O' Neil (Aberdeenshire) 'Kincardine of the O'Neils'. *See* Kincardine, above; the latter part of the name is to distinguish it from Kincardine in the Mearns, and refers to the monastery of Banchory–Ternan, founded by descendants of the Ui Néill clan of Ulster, and of which this parish was a property. Noted as Kincardyn Onele, c. 1200.

Kincraig (Highland) 'End of the crag'. *Ceann* (Gaelic) 'head', 'end', *na* 'of', *chreige* 'crag', 'rock'. Recorded in the seventeenth century as Kyncragye, but the *-ie* ending has been lost.

King Edward (Aberdeenshire) 'End of the (land) division'. *Cinn* (Gaelic) 'at the head of', *eadaradh* 'division'. Recorded as Kinedward, pre-1300. The name shows an attempt, when Gaelic was forgotten, to relate the sound to something apparently meaningful.

Kinghorn (Fife) 'At the head of the muddy ground'. *Cinn* (Gaelic) 'at the head of', *gronn* 'muddy land', 'marshland'. It was recorded as Kingorn in 1140.

Kinglass, Glen (Stirling) 'Glen of the dog stream'. The Gaelic form is Conghlais, from *con* 'dog', 'wolf', and *glas* 'water'. Noted as Kinglassin, 1224.

Kingskettle (Fife) The second part appears to be Gaelic or Pictish *cat*, 'cat', with Gaelic *-ail* indicating place, found mostly in Pictland. Old forms include Cattell, 1160, Kettil, 1558. 'Kings' indicating a royal property, appears to date from the 16th century: no early form is noted.

Kingston (Moray) Originally Kingston-upon-Spey, named after the English city of Kingston-upon-Hull by two expatriate timber exporters who set up shop here in 1784.

Kingussie (Highland) 'At the head of the pine wood'. *Cinn* (Gaelic) 'at the head of', *ghiuthasaich* 'abounding in pine trees'. Records show Kinguscy 1210, Kyngucy 1380.

Kinkell (Fife; Highland; other districts) 'Head of the wood', from Gaelic *ceann* 'head', *na* 'of', *coille* 'wood'. The Fife name is found as Kinnakelle, 1199.

Kinlochbervie (Highland) 'Head of Loch Bervie'. *Ceann* (Gaelic) 'head', 'end', Loch Biorbhaidh 'boiling or stormy loch'. See Inverbervie.

Kinlochewe (Highland) 'Head of Loch Ewe'. *Ceann* (Gaelic) 'head', 'end'; *Loch Iù* (Gaelic) possibly 'loch of the yew trees', though it may be a river name from a lost Pictish source. Kinlochewe is at the head of Loch Maree, which formerly bore the name Loch Ewe, noted as Loch Ew in Blaeu's *Atlas*, 1654. Kinlochewe remained as a 'fossil' name after the loch name changed. *See* Maree.

Kinlochleven (Highland) 'Head of Loch Leven'. *Ceann* (Gaelic) 'head', *loch* 'lake', 'loch', *léan* swampy place'. See Leven, Lomond.

Kinloss (Moray) Possibly 'head of the garden'. *Ceann* (Gaelic) 'head', *lios* 'garden'. Noted in 1187 as Kynloss. This may refer to the abbey founded here in 1151.

Kinnaird (Perth & Kinross; Aberdeenshire) 'Hill head'. *Ceann* (Gaelic) 'head', *àird* 'height'. Noted as Kinard, 1183.

Kinneil (West Lothian) 'Wall's end'. *Ceann* (Gaelic) 'head', 'end', *fhaill* 'wall's'. The location is close to the eastern end of the Antonine Wall. Its Pictish name was given by Bede in 731, *Peanfahel*; the *p*-Celtic form was subsequently displaced for Gaelic *cinn*-.

Kinross (Perth & Kinross) 'Head of the promontory'. *Ceann* (Gaelic) 'head', *ros* 'promontory'. The name was recorded as Kynros around 1144.

Kintail (Highland) 'Head of the sea-water'. *Ceann* (Gaelic) 'head', *an t-saille* 'of the salt-water inlet'. See Kentallen.

Kintore (Aberdeenshire) 'At the head of the hill'. *Ceann* (Gaelic) 'at the head of', *torr* 'steep hill'. Noted 1190 as Kynthor.

Kintyre (Argyll & Bute) 'Head of the land'. *Ceann* (Gaelic) 'head', *tire* 'of land'. A record of 807 indicates Ciuntire, 18th century Cantire.

Kippen (Stirling) 'Little hill'. *Ceap* (Gaelic) 'stumpy hill', 'block', *-an* (Gaelic diminutive suffix). Kippen is found in other names, like Kippendavie, near Dunblane, with the suffix *dabhach* (Gaelic), a unit of land measurement.

Kirk- Old Norse *kirkja* and Old English *cirice* both give Scots *kirk*. Many 'Kirk' names, especially Kirktons, are Scots, post-1400.

Kirkbister (Shetland) 'Church farm'. *Kirkja* (Old Norse) 'church', *bolstadr* 'farm'. Kirbister in Orkney has the same derivation.

Kirkcaldy (Fife) 'Fort on the hard hill'. *Caer* (Brittonic) 'fort', *caled* 'hard', *din* 'hill'. Records show Kirkaladunt from c1050, Kirkawde 1321.

Kirkconnel (Dumfries & Galloway) 'Connal's church'. *Kirk* (Scots from Old English *cirice* or Old Norse *kirkja*) 'church'; *Conall* (Old Irish proper name), a pupil of St Mungo. Recorded as Kyrkconwelle, 1347.

Kirkcudbright (Dumfries & Galloway) 'Church of St Cuthbert'. *Kirk* (Scots, from Old English *cirice*, or Old Norse *kirkja*) 'church'; *Cudberct* (Old English personal name meaning 'famous-bright'), the 7th-century ascetic St Cuthbert. Recorded as Kirkcutbrithe in 1291.

Kirkhope (Borders) 'Church in the valley'. *Kirk* (Scots, from Old Norse *kirkja*) 'church', *hop* 'enclosed valley'. Noted as Kyrchope, c. 1340. Hobkirk is the same, with reversed elements.

Kirkintilloch (East Dunbartonshire) 'Fort at the head of the hill'. *Caer* (Brittonic) 'fort'; *cinn* (Gaelic) 'at the head of', *tulaich* 'hill'. Found in the 10th century as Caerpentaloch; by around 1200 *pen* has given way to *cinn*, with Kirkintulach.

Kirk o' Shotts (North Lanarkshire) 'Church on the steep slopes'. *Kirk* (Scots, from Old English *cirice*, or Old Norse *kirkja*) 'church'; *sceots* (Old English) 'steep slopes'.

Kirkoswald (South Ayrshire) 'Church of St Oswald'. *Kirk* (Scots, from Old English *cirice*, or Old Norse *kirkja*) 'church'; *Oswald* (Old English personal name) the 7th-century king of Northumbria, who had links with this part of ancient Strathclyde.

Kirkpatrick (Dumfries & Galloway) 'Church of St Patrick'. There are several places of this name, and an additional name was sometimes supplied to make its identity clear, as in Kirkpatrick Fleming, 'Patrick's church of the Fleming(s)', from the 1300s.

Kirkwall (Orkney) 'Church on the bay'. *Kirkja* (Old Norse) 'church', and *vágr* 'bay'. Recorded in the *Orkneyinga Saga*, around 1225, as Kirkiuvagr, and later as Kirkvaw in a text of 1400. The terminal -ll is a Scots addition to make 'sense' of the Norse name.

Kirriemuir (Angus) 'The great quarter'. *Ceathramh* (Gaelic) a land measure that was a fourth of a *dabhach*, or 192 Scots acres, *mòr* 'great', 'big'. Noted as Kerimure, 1229.

Kishorn (Highland) 'Protruding cape'. *Keisa* (Old Norse) 'protrude', *horn* 'cape'. Recorded as Kischernis, 1464.

Kittybrewster (Aberdeen) 'Brewer's green'. *Cèide* (Gaelic) 'green', 'hillock'; *browster* (Scots) 'brewer'. Noted as Browster Lands, 1376.

Knock (Moray) 'Hill'. *Cnoc* (Gaelic) 'hill'. Both in its scotticised and Gaelic forms it occurs in many places, employed for lesser hills.

Knockan (Highland) 'Little hill'. *Cnocan* (Gaelic) 'small hill' from *cnoc* with diminutive *-an* suffix.

Knockando (Moray) 'Hill of the market'. *Cnoc* (Gaelic) 'hill', *cheannachd* 'of the market'. Found as Knockandoch, 1685.

Knoydart (Highland) Possibly 'Cnut's fiord'. *Cnut* (Old Norse personal name), *fjordr* 'firth', 'sea-loch'. Recorded as Knodworath, 1309. Its Gaelic name is *Cnoideart*.

Kyle (South Ayrshire; East Ayrshire) The old central division of Ayrshire. Although the name has been linked to the Brittonic king Coel Hen ('old King Cole', c. 400), the source may be from the Water of Coyle, which flows to join the Ayr, and whose name perhaps derives from Gaelic *caol*, 'narrow'. Recorded as Cyil by Bede, 731.

Kyleakin (Highland) 'Narrows of Haakon'. *Caol* (Gaelic) 'straits', 'narrows'; *Haakon* (Old Norse personal name). Several kings of Norway bore this name, though it may alternatively be called after a local magnate.

Kyle of Lochalsh (Highland) 'Narrows of Loch Alsh'. *Caol* (Gaelic) 'strait'. See Alsh, Loch.

Kylerhea (Highland) 'Reith's strait'. *Caol* (Gaelic) 'strait'; *Réithainn* (Old Gaelic personal name) in legend a warrior of giant size who jumped across the kyle.

Kylesku (Highland) 'Narrows of the strait'. *Caolas* (Gaelic) 'straits', *cumhann* 'narrow', 'thin'. The narrow sea entrance where Loch Cairnbawn ('white rock') meets the junction of Loch Glendhu ('dark glen') and Loch Glencoul ('glen of the nook').

L

Ladder Hills (Aberdeenshire) 'Hills of the slopes'. *Leitir* (Gaelic) 'hill slope'. 'Hills' is a later addition.

Ladybank (Fife) 'Boggy slope'. *Leathad* (Gaelic) 'slope', *bog* 'moist'.

Ladywell (West Lothian) 'Our Lady's Well', dedicated to the Virgin Mary. There are numerous other local Lady-names throughout the country, dating back to before the Reformation of 1560, mostly but not all with the same derivation (see Ladybank).

Laich of Moray (Moray) 'Lowland of Moray', from Scots *laich*, 'lowland', borrowed from Gaelic *leachd*, 'sloping ground'. See Menteith.

Laide (Highland) 'Slopes'. *Leathad* (Gaelic) 'slope'.

Lairg (Highland) 'The pass'. *Lairig* (Gaelic) 'pass', 'beaten path'. Noted as Larg, c. 1230.

Lamancha (Borders) A Spanish name bestowed by the proprietor, Admiral Cochrane, in the 1730s; formerly Grange of Romanno.

Lamlash (North Ayrshire) 'Isle of Mo-Laise'. *Eilean* (Gaelic) 'island', *Malaise* (Gaelic personal name, incorporating *mo* 'my', or 'my dear', and *Las* 'Flame'). Found as Almelasche, 1329. The 7th-century saint Mo-Laise's island gave its name to the village.

Lammermuir (Borders; East Lothian) Possibly 'lambs' moor'. *Lombor, lambre* (Old English) 'lamb'; *muir* (Scots from Old English *mor*) 'moorland'. Nowadays known as the Lammermuir Hills. An early 9th-century document records Lombormore, and a later text has Lambremor.

Lanark (South Lanarkshire) 'The glade'. *Llanerc* (Brittonic) 'forest glade'. Recorded as Lannarc 1188, Lanerch 1430.

Langholm (Dumfries & Galloway) 'Long water meadow'. *Lang* (Scots) 'long', *holm* (Scots from Old Norse *holmr*) 'water meadow', 'haugh'. Recorded in 1376 in this form.

Larbert (Falkirk) Possibly 'half wood'. *Lled* (Brittonic) 'half', 'part', *pert* 'wood'. Recorded as Lethberth 1195, Larbert 1251.

Largo (Fife) 'Steep place'. *Leargach* (Gaelic) 'steep slope'. The name may first have been applied to the hill of Largo Law. Recorded as Largaugh 1250, Largaw 1279, Largo 1595.

Largs (North Ayrshire) 'Hillside'. *Learg* (Gaelic) 'hillside'. -*s* ending of the name would appear to have been a later addition. It was documented Larghes around 1140.

Lasswade (Midlothian) 'Ford by the meadow'. *Leas* (Old English) 'meadow', *gewaed* 'ford'. A record of 1150 shows Leswade.

Latheron (Highland) 'Miry place'. *Làthach* (Gaelic) 'miry', seems likely, although the -ron ending is of unclear origin, perhaps the stream name (see Earn). In 1274 it was noted as Lagheryn. Latheronwheel nearby adds Gaelic *a' phuill* 'of the pool'.

Lauder (Borders) Takes its name from the Leader Water, which may be derived as *lou* (Brittonic) 'wash', and *dobhar* 'water'. *Lòthur* (Old Irish) 'trench' has also been suggested. Recorded as Louueder 1208, Lawedir 1250, Loweder 1298.

Laurencekirk (Aberdeenshire) 'St Laurence's kirk'. Founded by Lord Gardenstone in 1770 and at first named Kirkton of St Laurence. Previously it was Conveth, from Old Irish *coindmed* 'billeting', a place where the warriors of a chief were lodged with the people.

Lawers, Ben (Perth & Kinross) Probably 'Loud, resounding stream'. *Labhar* (Gaelic) 'loud'. The name was extended from the stream to Ben Lawers and the surrounding area. The Gaelic name is Beinn Labhair.

Laxford, Loch (Highland) 'Salmon fiord'. *Laks* (Old Norse) 'salmon', *fjord* 'sea-loch', 'firth'. Gaelic 'loch' was added in the post-Norse period.

Leadburn (Midlothian) 'Bernard's stone'. *Leac* (Gaelic) 'stone', *Bernard* (Old English personal name). The name is found as Lecbernard around 1200. In the Leadburn stream in South Lanarkshire, the Lead-element may be related to Leader/Lauder.

Leadhills (South Lanarkshire) The Scots name from around the 14th century reflects that here was an important site for the extraction of lead, as well as gold and silver.

Ledi, Ben (Stirling) Traditionally taken as 'mountain of God'. *Beinn* (Gaelic) 'mountain', *le*, 'in possession of', *Dia* 'God'. A more plausible if prosaic alternative is *leathad* (Gaelic) 'slope': 'hill of the sloping sides'.

Leith (Edinburgh) Possibly 'wet place'. *Lleith* (Brittonic) 'moist'. The Gaelic form is Lìte (*lì* and *lìth* 'water'). Records show Inverlet 1145, Leth 1570. Inverleith, with Gaelic *inbhir* 'river-mouth', remains a local district name.

Lennox (Stirling, East & West Dunbarton) Known as The Lennox, Gaelic Na Leamhanaich, the origin lies in the Brittonic name of the River Leven, meaning 'river of elms' and rendered into Gaelic as *leamhann*. The suffix -*ach*, gives it the sense of (place of) 'the folk of Lennox'.

Leny, River and **Pass** (Stirling) 'Narrow cattle path', from *lànaig* (Gaelic) 'narrow cattle path'. It was noted as Lani in 1237.

Lenzie (East Dunbartonshire) 'Wet meadow', from Gaelic *léanaidh* 'wet', 'marshy'. The *z*, as very often, is intended to convey the sound of Gaelic 'gh', a confusion caused by old writing style. Noted as Lenneth, c. 1230, and Lenye 1451.

Lerwick (Shetland) 'Mud bay'. *Leir* (Old Norse) 'mud', *vik* 'bay'. The name of the bay is Old Norse, but there was no town here until about 1600, long after the end of the Viking era.

Leslie (Fife; Aberdeenshire) Possibly 'garden by the pool'. *Lios* (Gaelic) 'enclosure,' 'garden', *linn* 'pool'. Both Leslies are located by streams. An alternative derivation, *llys* (Brittonic–Pictish) 'court' and *celyn* 'holly' has also been proposed. The Aberdeenshire name was recorded around 1180 as Lesslyn, in 1232 as Lescelin.

Lesmahagow (South Lanarkshire) Found in virtually this form, Lesmhagu, in 1138, suggesting the first part to be from Gaelic *lios*, 'enclosure', 'garden'. Found also as Ecclesia Machuti, 'church of Mahagow' in 1148. Mahagow may stem from Gaelic *Mo-Fhegu* ('my Fechin'), making it 'St Fechin's enclosure'.

Letham (Angus; Fife; Stirling) This has been derived as 'village of the barns' from *hlatha* (Old English) 'barns', and *ham* (Old English) 'village'. But perhaps Gaelic *leathan* 'broad slope' is more likely.

Letters (Highland) 'Hill slopes'. *Leitir* (Gaelic) 'hillside'.

Leuchars (Fife) Probably 'place of the rushes'. *Luachair* (Gaelic) 'rushes'. Noted pre-1300 as Locres.

Leven (Fife; West Dunbartonshire; Argyll & Bute) Probably 'elm river'. The Fife name, noted as Lochleuine around 955, is from the local River Leven, derived from *leamhain* (Gaelic) 'elm'. In the case of the River Leven flowing from Loch Lomond, noted as Lemn in the 9th-century, however, the indication by Ptolemy, c. AD 150, of a Lemannonios Kolpos (gulf), pre-dates Gaelic names in Scotland and may be a Brittonic 'elm river'.

Leven, Loch (Highland) The West Highland Loch Leven also takes its name from its river, perhaps with the same Gaelic derivation as above though in this case *lèan* 'swampy place', has also been suggested. Noted as Glen Lemnae, 704.

Leverburgh (Western Isles) A modern name joining the family name of Lord Leverhulme, proprietor of Lewis in the 1920s, with *burgh* (Scots) 'town'. The village was formerly Obbe (Gaelic) 'bay').

Lewis (Western Isles) The first element is unclear, possibly an Old Norse adaptation of an earlier name. 'Song house' from *Ljod* (Old Norse) 'song'; *hus* 'house' has been suggested. In Gaelic it became *Leòdhas*, and has been confused with *leoghuis* (Gaelic) 'marshiness'. Recorded as Leodus and Lyodus c. 1100, Liodhus in the *Orkneyinga Saga* c. 1225, Leoghuis 1449.

Lhanbryde (Moray) 'Church-place of St Bride'. The form of the prefix is perhaps unique in Scotland. *Làn* is an obsolete Gaelic word for 'church'; its later meaning of 'field' often meant a field belonging to a church. Older forms include Lamanbride in 1215, Lambride, late 14th century.

Liberton (Edinburgh) Once thought to be 'Lepers' place': *Liber* (Old English) 'leper', *tūn* 'place'. More probably it is 'hillside barley farm' from Old English *hlith* 'hill', *bere* 'barley', and *tūn* 'farmstead'. Found as Liberton in 1128.

Liddesdale (Borders) 'Dale of the Liddel Water'. A tautological name, as Liddel incorporates both Old Norse *hlyde* 'noisy', and *dalr* 'dale'. Recorded as Lidelesdale, 1179.

Liff (Angus) Perhaps '(place of) herbs'. *Luibh* (Gaelic) 'plant', 'herb'. Recorded in this form c. 1120. *See* Luce, Luss.

Lindores (Fife) 'Field or lake of the pass'. Either *lann* (Gaelic) 'field', or *llyn* (Brittonic) 'lake'; *dorus* (Gaelic) 'door', 'entryway' (see Durrisdeer). Noted c. 1182 as Lundors.

Linlithgow (West Lothian) 'Place by 'the lake in the moist hollow'. *Llyn* (Brittonic) 'lake', *lleith* 'moist', and *cau* 'hollow'. Recorded in 1124 as Linlitcu.

Linnhe, Loch (Highland) 'The pool'. *Linne* (Gaelic) 'pool'. The seaward end, past the Corran narrows, is known in Gaelic as An Linne Sealach, *sealach* explained variously as 'salty', and 'of the willows'. The inner loch is An Linne Dubh, *dubh* (Gaelic) 'black'.

Linton, East and West (East Lothian) 'Homestead by the water'. *Linn* (Brittonic) 'pool'; *tūn* (Old English) 'farmstead'. Gaelic *líon*, 'flax' is also suggested, for 'flax farm'. Noted in 1127 as Lintun.

Linwood (Renfrewshire) 'Wood by the pool'. *Llyn* (Brittonic) 'pool'; *wudu* (Old English) 'wood'.

Lionel (Lewis) 'Flax hill'. *Líon* (Gaelic) 'flax'; *hóll* (Old Norse) 'hill'.

Lismore (Argyll & Bute) 'Big garden'. *Lios* (Gaelic) 'garden', 'enclosure', *mòr* 'big'.

Livingston (West Lothian) 'Leving's place'. *Leving* (Old English personal name), *tūn* 'farmstead'. Recorded as Uilla Leuing, 1124, Leuinistun 1250.

Lix (Highland; Angus; Perth & Kinross) 'Flagstones'. *Lic* (Gaelic *leac* in the locative form) 'flagstone'.

Loanhead (Midlothian) 'At the top of the lane'. *Loan* (Scots) 'lane', originally a path to the *loaning* (Scots) 'enclosed pasture land'. Recorded as Loneheid in 1618.

Loch See under specific names, Achray, etc., except where the names have been run into one. There is a Loch Loch below Beinn a'Ghlo in Perthshire; the second element is Old Gaelic *lòch*, 'dark'.

Lochaber (Highland) Probably 'area of the loch confluence'. *Loch* (Gaelic) 'sea-loch', 'lake'; *aber* (Pictish-Brittonic) 'at the confluence of'. Noted c. 700 as Stagnum Aporum, 'swamp of the confluences'; in 1297 as Lochabor.

Locharbriggs (Dumfries & Galloway) 'Reed stacks'. Lochar from *luachar* (Gaelic) 'rushes', presumably referring to the reedy moss through which the Lochar river runs. 'Briggs' has been assumed to be Scots *brig*, 'bridge', but the plural form suggests the plural of *brig* (Gaelic) 'heap', 'pile', referring to gathered and piled rushes.

Lochboisdale (Western Isles) Gaelic *loch* has been added to a name already indicative of a coastal feature: *bug* (Old Norse) 'bay'. The termination is *dalr* 'dale'.

Lochcarron (Highland) Formerly known as Jeantown, this village takes its name from the loch on which it is situated. *See* Carron.

Lochee (Dundee) 'Corn loch'. *Loch* (Gaelic) 'lake', 'loch', *iodh* 'corn'.

Lochgelly (Fife) Place of 'the shining loch'. *Loch* (Gaelic) 'loch', 'lake', *geal* 'bright', 'shining'.

Lochgilphead (Argyll & Bute) 'Head of Loch Gilp'. *Loch* (Gaelic) 'sea-loch', *gilb* 'chisel'. The name of the loch, noted as Louchgilp, c. 1246, indicates its shape.

Lochinvar (Dumfries & Galloway) 'Loch of the height'. *Loch* (Gaelic) 'lake', 'loch', *an* 'of the', *bharra* 'height'. Found in this form 1540.

Lochinver (Highland) 'Loch at the river mouth'. *Loch* (Gaelic) 'lake', 'loch', *inbhir* 'river-mouth'.

Lochmaben (Dumfries & Galloway) Mapon was a Celtic deity, and 'Mapon's loch' would be apt for what seems a cult site. An alternative is 'loch by the bare-topped hill'. *Loch* (Gaelic) 'loch', 'lake', *maol* 'bare top', *beinn* 'hill'. The Latin form was Locus Maponis. Noted as Locmaban, 1166.

Lochmaddy (Western Isles) 'Loch of the Dog'. *Loch* (Gaelic) 'lake', 'loch', *nam* 'of the', *mhadaidh* 'dog'. *See also* Portavadie.

Lochnagar (Aberdeenshire) 'Loch of the noise, or laughter'. *Loch* (Gaelic) 'loch', 'lake', *na* 'of the', if from *gàir* 'noise'; if from *gàire* 'laughter'. The mountain name is from the loch at its foot, noted as Loch Garr in 1640. It was also known as Beinn nan Ciochan (Gaelic) 'the mountain of the paps, or breasts'. Prudishness may have encouraged the alternative name.

Lochore (Fife) Possibly 'brown loch'. *Loch* (Gaelic) 'lake', 'loch', *odhar* 'brown'; perhaps referring to the peaty soil of the area. But, as the name comes from the River Ore, it may be cognate with that of the Ayr and the English Ore and Oare, from a conjectural pre-Celtic root-form *ora, indicating 'flowing'. Found as Lochor, 1241.

Lochty (Fife; Perthshire; Angus; Moray) 'Stream of the black goddess'. *Loch* here seems to be from *lòch* (Old Gaelic) 'dark'; with *dae* (Irish Gaelic) 'goddess', related to Gaelic *dia*. The reference is to a 'black goddess' river spirit. See also Lochy, Munlochy.

Lochtyloch (West Lothian) 'Dark hill'. The name stems from *lòch* (Old Gaelic) 'dark', and *tulach* (Gaelic) 'hill'.

Lochwinnoch (Renfrewshire) Loch of St Wynnin or
Finnian. Noted as Lochynoc, 1158. See Kilwinning.

Lochy, River, Loch, Glen (Highland; Argyll & Bute;
Moray) 'Stream of the black goddess'. *Lòch* (Old Gaelic)
'black'; *dae* (Irish Gaelic) 'goddess', related to Gaelic *dia*.
The name comes originally from the river in each case.
The River Lochy in the Great Glen is referred to as *dea
nigra*, 'black goddess', and its loch as Lacus Lochdiae, by
Adamnan, c. 700.

Lockerbie (Dumfries & Galloway) 'Lokard's village'.
Lokard (Old Norse personal name); *by* 'village',
'farmstead'. Recorded as Lokardebi in 1306.

Logie (Highland; elsewhere) 'Place in the hollow'. *Lagaigh*
(Gaelic) 'in the hollow'. Found as Logyne, 1184; the
terminal -*n* explained as 'a scribe's flourish'.

Logierait (Perth & Kinross) 'My-Coeddi's hollow'. *Lagaigh*
(Gaelic) 'in the hollow', *mo* 'my'; *Choid* (Irish Gaelic
proper name). Coeddi was a bishop of Iona (died 712).

Lomond, Loch and **Ben** (Stirling; Fife) One possible
source is Brittonic *llumon*, 'beacon', as in the Welsh
mountain Pumlumon. This suits Ben Lomond, Beinn
Laomuinn in Gaelic, and the twin Lomond Hills (Fife)
whose positions make them suitable as beacon hills. Also
possible is *leamhan* (Gaelic) 'elm'. This accounts for the
name of the River Leven which flows from Loch
Lomond to the Firth of Clyde. The 9th-century writer
Nennius wrote of 'the great lake Lummonu, which in
English is called Lochleven, in the region of the Picts';
and it is documented in a text of 1535 as Levin. It may be
that for a long period the loch was referred to as both
Lomond and Leven. The Gaelic form is Loch
Laomuinn, and possibly – if the mountain name Beinn
Laomuinn is from *llumon* – the similarity of the names
caused the loch name to be altered from its river's name
to that of the beacon-mountain which rises above it. It is
notable that the Fife Lomonds also rise above a Loch
Leven.

Long, Loch (Argyll & Bute; Highland) 'Loch of ships'. *Loch* (Gaelic) 'lake', 'loch', *luing* 'of ships'. Noted as Loch Long 1225, but it has been identified with Ptolemy's pre-Gaelic Lemannonios Kolpos (see Leven).

Longannet (Fife) 'Field of the patron saint's church'. *Lann* (Gaelic) 'field', *annat* 'patron saint's church', or 'church with relics' (see Annat).

Longart (Highland) 'Camping place'. *Long* (Irish Gaelic) 'ship', came also to mean 'dwelling'; and *phort* (Irish Gaelic) 'harbour', came also to mean 'encampment'. Luncarty near Perth, noted as Lumphortyn, 1250, and in Aberdeenshire also have the 'camp' meaning.

Longforgan (Perth & Kinross) 'Field, or church, over the boggy place'. *Lann* (Gaelic) means both 'field' and 'church'; *for* 'over, above'; *gronn* (Old Gaelic) 'marsh'. It appears as Langforgrunde in the 14th century.

Longformacus (Borders) Apparently 'church on the land of Maccus'. Long (Brittonic *lann*, cognate with Welsh *llan*, 'church'), *fothir* 'land, meadow'; *Maccus* (Irish-Scandinavian form of personal name Magnus), seen also in Maxwell. Recorded as Langeford Makhous c. 1340.

Longhope (Orkney) 'Long sheltered bay'. *Hop* (Old Norse) 'sheltered bay'.

Longniddry (East Lothian) 'Church of the new hamlet'. Brittonic *lann* 'church', *nuadh* 'new'; *tref* 'hamlet'. Recorded as Langnedre in 1595.

Lorn(e) (Argyll & Bute) This area (Lathurna in Gaelic), together with the firth of the same name, was named after Loarn, brother of Fergus of Ulster, and by tradition one of the leaders of the Scots' colonisation from Ireland. Noted as Lorne in 1304.

Lossie, River (Moray). The river-name has been linked to the name Loxa on Ptolemy's 2nd-century map, from a Greek root *loxos*, 'crooked'. A derivation from *lus* (Gaelic) 'herbs', 'plants,' has also been suggested. The name Lossiemouth indicates that the town was only developed in the late 17th century, when a harbour was built here.

Lothian This area, once a kingdom of the Britons, later part of the Anglian kingdom of Bernicia, is believed to have been named after its historical founder, Leudonus (Brittonic or pre-Celtic personal name) of uncertain origin. Early records show Loonia c. 970, Lothene 1091, Louthion c. 1200, Laodinia 1245.

Loudoun (East Ayrshire) 'Beacon hill'. *Lowe* (Scots 'fire' from Old Norse *logr*) and *dún* (Old English) 'hill'. An association with the Celtic god Lug has also been suggested, making the name *Lugdunon*, cognate with Lyons in France (*Lugdunum*). Found in its present form c. 1140.

Loyal, Ben and **Loch** (Highland) An anglicised version of the Gaelic Beinn Laoghal, from *beinn* (Gaelic) 'mountain'; *laga* (Old Norse) 'law', and *fjall* 'hill'. This would indicate a meeting-place. But *leidh* (Old Norse) 'levy' or 'mustering-place' has also been suggested. Noted as Ben Lyoll, 1601.

Lubnaig, Loch (Stirling) 'Loch of the bend'. *Lùb* (Gaelic) 'bend', with a double suffix, -*an*, and -*aig*, both indicating 'small'.

Luce (Dumfries & Galloway) 'Place of herbs or plants'. *Lus* (Gaelic) 'herbs', 'plants'. Noted as Glenlus, 1220. *See* Luss.

Lugar, River (East Ayrshire) 'Bright stream'. A Brittonic name conjectured from the early Celtic root form **loucos* 'white', with the termination -*ar*, indicating a river. The name of the Celtic god Lugh comes from the same source, and it could be 'stream of Lugh'. Found in this form c. 1200. The River Luggie shares the derivation.

Lui, Ben (Stirling) 'Mountain of calves'. *Beinn* (Gaelic) 'mountain', *laoigh* 'of calves'. See Ardlui.

Luichart, Loch (Highland) 'Place of encampment'. *Long* (Irish Gaelic) 'dwelling', 'ship'; *phort* (Irish Gaelic) 'harbour', 'camp ground'. Longphort is compressed into Luichart. *See* Longart. Gaelic *lùchairt* 'palace' has also been suggested, but this seems much less likely.

Luing (Argyll & Bute) 'Ship island'. *Luing* (Gaelic) 'ship's.'

Lumphanan (Aberdeenshire) 'Finnan's field'. *Lann* (Gaelic) 'field', 'enclosure'; *Fhìonain* (Gaelic proper name) 'Finnan's.' The site of a property belonging to a church dedicated to St Finnan. Lumphinnans in Fife is likely to have the same derivation.

Lunan (Angus) Perhaps 'stream of health' from Brittonic *lūn* 'healthy', 'pure', from the same stem as Gaelic *slàinte* 'health'. 'Wave bay' from *lunnan* (Gaelic) 'sea waves' has been proposed; if so, the name has worked its way up the Lunan Burn from the estuary, noted as Innirlunan, 1189, to Lunanhead, beyond Forfar. There is another Lunan Burn flowing into the River Isla, far from the sea.

Lundin Links (Fife) Perhaps 'boggy site'. A Pictish name which may be associated with *lodan* (Gaelic) 'marsh'. Noted as Lundin around 1200. The addition of Links in the post-Gaelic period is from *hlinc* (Old English) 'rising ground', 'bank', used in Scots to describe grassy dunes by the sea, and by association, golf courses.

Luss (Argyll & Bute) 'Place of herbs or plants'. *Lus* (Gaelic) 'herbs', 'plants'. Recorded as Lus, 1225. *See* Luce.

Lybster (Highland) 'Settlement in the lee'. *Hlie* (Old Norse) 'leeward', *bolstadr* 'farmstead', 'settlement'.

Lyne Water (Fife; Borders) The two Lynes have different derivations; that in Fife is from *lleith* (Brittonic) 'moist', 'wet', cognate with Leith, and noted as aqua de Letheni, 'Leithen water', 1227; that near Peebles is Gaelic *linne* 'pool, 'waterfall', noted as Lyn, c. 1190.

Lyon, River, Glen, Loch (Perth & Kinross) Apparently 'grinding river'. The Gaelic name is *Lìobhunn*, deriving from a pre-Celtic root **lim* 'file', presumably with reference to the erosive action of the river.

M

Macdui, Ben (Moray) 'Hill of the sons of Dubh or Duff'. *Beinn* (Gaelic) 'mountain', *mac Duibh* 'sons of Duff'. The spelling Macdhui is also found.

Macduff (Aberdeenshire) Named in 1783 by James Duff, second earl of Fife, who redeveloped the settlement, previously known as Down (Gaelic *dùn* 'fort'). Macduff as a surname simply means 'son of *Dubh*, the black-haired one'.

Machars, The (Dumfries & Galloway) 'Plains'. *Machair* (Gaelic) 'low-lying fertile plain', most familiarly, the grassy strip just inland from a beach.

Machrihanish (Argyll & Bute) 'Coastal plain of Sanas'. *Machair* (Gaelic) 'low-lying fertile plain'. The latter part may be a personal or district name. It has also been linked to *sean-innse* (Gaelic) 'old haugh or water-meadow'.

Macmerry (East Lothian) The first part is 'plain', from *magh* (Gaelic) 'plain'; the second part is perhaps related to the Gaelic root *mear-*, indicating 'exposed', though *mear* also means 'merry'.

Maddiston (West Lothian) 'Mandred's place'. A reference of 1366 has Mandredestone; *Mandred* (Old English personal name), *tūn* 'settlement'.

Maggieknockater (Aberdeenshire) 'The fuller's plain'. *Magh* (Gaelic) 'plain', *an* 'of the', *fhucadair* 'fuller'.

Mainland (Orkney; Shetland) The name for the largest island of both groups is of some antiquity, from Old Norse *megin*, 'principal', *land*, 'land'; found as Meginland c. 1150.

Mains (All regions) Very often found attached to a farm name, indicating the main or home farm of an estate. It is an aphetic form (losing the unstressed first sound) of the word 'domain', noted from 1479.

Mallaig (Highland) Possibly 'headland bay'. *Muli* (Old Norse) 'headland'; *aig* (Gaelic version of original Old Norse *vágr*) 'bay'. An alternative derivation of the first part is from *mol* (Old Norse) 'shingle'.

Mamore (Highland) 'Big round hills'. *Màm* (Gaelic) 'rounded hill', *mòr* 'big'. The converse form of Mambeg (from Gaelic *beag*, 'small') is found in several places.

Manuel (West Lothian) Probably 'rock of the view'. *Maen* (Brittonic) 'rock', *gwel* 'view', 'outlook'. Noted as Manuell, c. 1190.

Mar (Aberdeenshire) Mar with Buchan was one of the seven divisions of ancient Pictland. It may be an unidentified personal name. The Gaelic form is Marr, noted in the *Book of Deer*, c. 1150.

Marchmont (Edinburgh) 'Horse hill'. *Marc* (Gaelic) 'horse', *monadh* 'hill'.

Maree, Loch (Highland) 'Maelrubha's loch'. Until the 17th century it was known as Loch Ewe, from the river flowing from its western end. 'Maree', found as Loch Maroy, 1638, denotes the missionary saint Maelrubha. An island in the loch hosted a pre-Christian cult which was transferred to him, becoming known as Eilean Maruibhe. In time this name was extended to the loch itself. See Kinlochewe.

Markinch (Fife) 'Isle or water meadow of the horse'. *Marc* (Gaelic) 'horse', *innse* 'water meadow', 'island'. It was recorded as Marcinche around 1200.

Maryculter (Aberdeen) 'Back land of the (church of) Mary'. *See* Peterculter.

Maryhill (Glasgow) Named in 1760 after the local landowner, Mary Hill of Gairbraid.

Mauchline (East Ayrshire) 'Plain with a pool'. *Magh* (Gaelic) 'plain', *linne* 'pool'. Noted c. 1000 as Machlind i Cuil, 'Mauchline of the nook'.

Maud (Aberdeenshire) 'Dog's, or wolf's place'. *Madadh* (Gaelic) 'dog', 'wolf'.

Mawcarse (Perth & Kinross) 'Plain of the carse'. *Magh* (Gaelic) 'plain'; *carse* (Scots, from Old Norse *kerss*) 'low-lying river bank'.

May, Isle of (Midlothian) 'Isle of seagulls'. *Má* (Old Norse) 'seagull', *ey* 'island'. In the *Orkneyinga Saga* (c. 1225) it is referred to as Maeyar.

Maybole (South Ayrshire) 'House on the marsh'. Old forms are Maiboile, Minyybole, suggesting either Gaelic *magh* 'plain' or *'moíne* 'moss', 'bog' with *both* 'hut'. An old local rhyme says "It sits aboon a mire."

Mearns, The (Aberdeenshire, East Renfrewshire) Explained as 'the stewardship'; in Gaelic An mhaoirne, indicating an area administered by an officially appointed steward. But the terminal -*s* is not explained. The Mearns district south of Glasgow may stem from *magh* (Gaelic) 'plain', with the river-name Earn (see Earn).

Meigle (Perth & Kinross) 'Swampy field'. *Mig* (Brittonic–Pictish) 'swamp', *dol* 'meadow'. Noted as Miggil, 1183.

Meldrum (Aberdeenshire) The modern form suggests 'mountain ridge'; *meall* (Gaelic) 'mountain', *druim* 'ridge', but early forms, Melgedrom 1291, and Melkidrum 1296, make this origin improbable. The meaning of the prefix is uncertain, perhaps *melg* (Old Gaelic) 'milk'.

Melrose (Borders) 'Bare moor'. *Mailo* (Brittonic) 'bare', *ros* 'wood', 'moor'. Recorded as Mailros c. 700.

Melvich (Highland) 'Bay of sea-bent dunes'. *Mealbhan* (Gaelic) 'sea-bent', from *melr* (Old Norse) 'grassy dune', *vik* 'bay'. Melvaig has the same derivation.

Menstrie (Clackmannanshire) 'Hamlet in the plain'. *Maes* (Brittonic) 'open field', 'plain', *tref* 'settlement', 'hamlet'. Found as Mestryn in 1261.

Menteith, Lake of (Stirling) 'Lowland of Menteith', from
  leachd (Gaelic) 'sloping ground'. The Gaelic form,
  Leachd Teàdhaich, omitting the prefix but scotticised as
  'laich of Menteith', may explain the enduring insistence
  on referring to this loch as Scotland's only 'lake'.
  Menteith may be from *mòine* (Gaelic) 'peat moss', 'bog',
  or *mon*, a local form of *monadh* 'hill', with Teith (Celtic
  river-name). Early maps show Loch Monteith in the
  Laicht of Monteith. Recorded as Menetethe 1185;
  Monteath 1724. See Teith.
Merchiston (Edinburgh) 'Merchion's farm'. *Merchiaun*
  (Brittonic personal name); *tūn* (Old English) 'farmstead'.
Merkland (Dumfries & Galloway; other regions) 'Land held
  for the rental of one merk'. Scots *merk*, 'mark', a unit of
  currency. A common locality name, especially in the
  south-west.
Merrick (Dumfries & Galloway) 'Pronged hill'. *Meurach*
  (Gaelic) 'pronged', 'branchy'. Noted as Maerach Hill on
  Blaeu's map, 1654.
Merse, The (Borders) 'Lowland', from *maersc* (Old
  English) 'marsh', which came to mean 'low, flat land' in
  Scots. Noted in this form, 1560.
Methil (Fife) 'Middle church'. Early forms from 1220 have
  Methkill: Gaelic *meadhon*, 'middle', *cill* 'church'. Less
  probably 'wood boundary'. *Maid* (Brittonic) 'boundary';
  choille (Gaelic) 'of the wood'.
Methven (Perth & Kinross) 'Middle stone'. *Meddfaen*
  (Brittonic) 'middle stone', 'middle marker', as of a
  boundary. Noted as Methfen, 1211.
Milngavie (East Dunbartonshire) Perhaps 'windmill'.
  Muilleann (Gaelic) 'mill', *gaoithe* 'wind'. Windmills were
  unusual in Scotland, where water power was plentiful.
  Alternatively, it may be *Meal-na-gaoithe* (Gaelic) 'hill of
  the wind'. Noted on Blaeu's map, 1654, as Milguy.
Millport (North Ayrshire) Named after the large grain mill
  that stood above the harbour when the town was
  developed in the first decade of the nineteenth century.

Milton (All regions) 'Place of the mill'. Often a Scots version of the original *Baile a' Mhuileann* (Gaelic) 'place of the mill', found in anglicised form as Balavoulin.

Minch, The (Highland, Western Isles) Possibly 'great headland(s)'. *Megin* (Old Norse) 'great', *nes* 'headland'. The name is almost certainly Scandinavian in origin. The cape in question could either be Cape Wrath, or the Butt of Lewis, or both. The Gaelic name is Cuan nan Orc, 'sea of whales'; for the Little Minch, Cuan Sgithe, 'sea of Skye'.

Mingulay (Western Isles) Probably 'big island'. *Mikla* (Old Norse) 'big', has become transposed to *mingil*, with *ey* 'island'.

Moffat (Dumfries & Galloway) Place of the 'long plain', *Magh* (Gaelic) 'plain', *fada* 'long'. Noted as Moffet, 1179.

Moidart (Highland) Place of the 'muddy fiord'. *Moda* (Old Norse) 'mud'; *art* (Gaelic adaptation of Old Norse *fjordr*) 'sea-loch'. Recorded as Muddeward, 1292.

Monadhliath (Highland) 'Grey mountains'. *Monadh* (Gaelic) 'mountain(s)', from Celtic **monid* 'hilly ground', *liath* 'grey'.

Moniaive (Dumfries & Galloway) Possibly 'moor of crying'. *Moine* (Gaelic) 'moor', 'peat bog', *èibhe* 'cry', 'death-cry'. Recorded in 1560 as Monyyife.

Monifieth (Dundee) 'Peat-bed of the bog'. *Moine* (Gaelic) 'peat bed', *feithe* 'bog'. Recorded as Munifieth 1178.

Monklands (North Lanarkshire) 'The monks' lands'. In the 12th century King Malcolm IV granted lands here to the monks of Newbattle Abbey.

Montrose (Angus) 'The peat-moss of the promontory'. *Moine* (Gaelic) 'moor', peat bed', *ros* 'promontory'. The *t* in the name has been interpolated. Records show Munros c. 1200, Montrose 1296, Monros 1322, Montross 1480.

Monymusk (Aberdeenshire) 'Mucky peat bog'. Gaelic *moine*, 'moor, peat bed', *mosach* 'foul.'

Monzievaird (Perth & Kinross) Probably 'cornland of the bard'. *Magh* (Gaelic) 'plain', *an* 'of', *eadha* 'corn', *bhàrd* 'bard's'. Recorded as Muithauard, c. 1200. Bards were frequently given grants of land: the Gaelic form *-bhaird*, 'bard's', is often found incorporated in local names.

Moorfoot Hills (Midlothian) 'Moor place', from Old Norse/Old English *mór*, 'moor', and *thweit*, 'place', with the sense of a place cleared for grazing, one of the few instances of a mainland 'thwaite'. Noted as Morthwait, c. 1142.

Morar (Highland) 'Big water'. *Mór* (Gaelic) 'big', *dobhar* (Gaelic) 'water'. Noted as Morderer c. 1292.

Moray 'Sea settlement'. *Mori* (Old Gaelic related to Brittonic–Pictish *mor-tref*, 'sea-home'). Noted in the *Pictish Chronicle*, c. 970, as Moreb; latinised into Moravia, 1124.

More, Ben (several locations) 'Big mountain'. *Beinn* (Gaelic) 'mountain', *mòr* 'big'. The best-known Ben More is probably that above Crianlarich; some are identified by a locative name, as in Ben More Assynt, Ben More Mull, etc.

More, Glen (Highland) 'Great glen'. *Gleann* (Gaelic) 'glen', *mòr* 'big'. Known also as The Great Glen and Glen Albyn (Gaelic *Albainn*, 'of Scotland').

Moriston, River and **Glen** (Highland) 'Big waterfalls'. *Mòr* (Gaelic) 'big', *easain* 'waterfalls'. The *t* is a recent intrusion into the name; 19th century forms show Glen Morison.

Mormond (Aberdeenshire) 'Big hill'. *Mór* (Gaelic) 'big', *monadh* 'hill'.

Morningside (Edinburgh; Lanarkshire) Possibly 'Morgan's seat', but no early record. Most likely a 17th century name for a south-facing estate.

Morton (Dumfries & Galloway; Fife; Renfrewshire) 'Farm by the moor'. *Muir* (Scots) 'moor', *toun* from Old English *tūn*, 'farm'.

Morven (Aberdeenshire; Highland) 'Big hill'. Usually taken to be a transposition of Ben More, from *mòr* (Gaelic) 'big' and *beinn* (Gaelic) 'mountain'. However, if its source is Brittonic–Pictish *morwen* 'maiden' – and both Morvens are in Pictland – it is likely to be yet another mountain name drawing on the resemblance of its shape to the female breast.

Morvern (Argyll & Bute) 'Sea gap'. *Mor* (Old Gaelic) 'sea'; *bhearn* (Gaelic) 'gap'. Noted in 1343 as Gawrmorwarne, with the Gaelic prefix *garbh*, 'rough'. In Gaelic it is A'Mhorbhairn.

Motherwell (North Lanarkshire) 'The Mother's well'; Modyrwaile in 1363. The reference is to an ancient well dedicated to the Virgin Mary.

Moulin (Perth & Kinross) 'Bare hill'. *Maolinn* (Gaelic) 'bleak hill-brow'. Found as Molin, 1207.

Mount Vernon (Glasgow) Once known as Windyedge, said to have been renamed by George Buchanan, one of Glasgow's 18th-century 'tobacco lords', after the Washington family's plantation in Virginia.

Mounth (Angus; Aberdeenshire; Perth & Kinross) 'The mountain(s)' from *monadh* (Gaelic) 'mountain'. The original Scots name of the mountains miscalled 'Grampians', recorded as Muneth, 1198.

Mousa (Shetland) This island name is of uncertain derivation, though *mose* (Old Norse) 'moss', has been suggested as it was a source of peat.

Moy (Highland) 'The plain'. From *magh* (Gaelic) 'plain'. Noted as Muy, c. 1235.

Muck (Highland) 'Pig island'. *Muc* (Gaelic) 'pig'. The reference is usually taken to pigs being kept here rather than a topographical or totemic feature.

Muckersie (Perth & Kinross) 'Pigs' bank'. *Muc* (Gaelic) 'pig'; *kerss* (Old Norse) 'low-lying river bank' – the source of Scots *carse*.

Muckhart (Clackmannanshire) 'Pig yard'. *Muc* (Gaelic) 'pig', *gart* 'yard', 'enclosure'. Noted as Mukard, 1250.

Muckle Flugga (Shetland) 'Great cliffs'. *Micil* (Old Norse) 'great', 'big', *fluga* 'cliffs'.

Muir of Ord (Highland) 'The moor of the rounded hill'. *Muir* (Scots from Old English *mór*) 'moor'; *ord* (Gaelic) 'rounded hill'. The Gaelic name is Am Blàr Dubh, 'at the black field'.

Muirkirk (East Ayrshire) 'Church on the moor'. *Muir* (Scots from Old English *mor*) 'moor', *kirk* (Scots, from Old English *cirice*) 'church'.

Mulben (Moray) 'Bare hill'. *Maol* (Gaelic) 'bare, bald', *beinn* 'mountain'. Noted as Molben c. 1328.

Mull (Argyll & Bute) In Ptolemy's map CAD 150, it may be the island referred to as Maleos, making an Old Norse derivation, 'island of the headland' from *muli* 'headland', impossible. Other suggestions include *maol* (Gaelic) which as *maolinnn* can mean 'rocky brow' as well as 'bare summit'; and *meuilach* (Gaelic) 'favoured one'. The Gaelic name today is Muile, or Eilean Muileach. A Celtic source seems most likely, perhaps an early form of *maol*.

Munlochy (Highland) 'At the foot of the black goddess's stream'. *Bonn* (Gaelic) 'foot'; *lòch* (Old Gaelic) 'black'; *dae* 'goddess', giving Gaelic *dia*). Its Gaelic name now is Poll Lochaidh, 'black mud pool'.

Murrayfield (Edinburgh) Named after an eighteenth-century local landowner, Archibald Murray.

Murthly (Perth & Kinross) 'Big hill'. *Mòr* (Gaelic) 'big', *tulach* 'hill', possibly an artificial mound.

Musselburgh (East Lothian) 'Mussel town'. *Musle* (Old English) 'mussel', *burh* 'town'. Documented in 1000 as Musleburge.

N

Nairn (Highland) Originally the river name, pre-Celtic, from a conjectured Indo-European root *ner-* with the sense of 'penetrating' or 'submerging'. The town was at times known as Invernairn, Invernaren 1189, Narne 1583, Narden 17th century. The Gaelic form is Narrun.

Naver, River, Loch, Strath (Highland) Noted on Ptolemy's map of c. AD 150 as Nabaros, the name derives from a pre-Celtic root which has been identified both as *nebh*, indicating 'water' (*see* Nevis), and *nabh*, indicating 'fog' or 'cloud'; with the Celtic *-ar* ending, indicating a river.

Navitie (Fife; other areas) 'Holy place'. From Gaelic *neimheidh*, 'sacred or holy place', a term linked to Gaulish *nemeton*, 'place for ritual meetings'. The sense was adopted into Christianity as 'land belonging to the church'. Navidale in east Sutherland shows how Norse settlers absorbed the Gaelic term into a hybrid with Old Norse *dalr*, 'dale, valley'.

Ness, River, Loch (Highland) The name was originally that of the river; of unknown, probably pre-Celtic origin. In Adamnan's *Life of St Columba* it is referred to by the latinised *Nesa*. It long predates any likelihood of connection with Old Norse *nes*, 'cape.' In Gaelic it is Nis.

Netherby (Dumfries & Galloway) 'Lower farm'. *Nedri* (Old Norse) 'lower', *by* 'farmstead'.

Nethy, River (Perth & Kinross; Highland) 'Shining stream'. *See* Abernethy.

Nevis, River, Ben, Loch (Highland) Its origin is unclear; probably first that of the river, perhaps from the pre-Celtic root form **nebh*, 'cloud', as has been suggested for Naver. A case has also been made for *nimheis* (Old Gaelic) 'venomous'. Many other unattested interpretations have been made. The form Nevess is found from 1552. Its Gaelic form is Beinn Nibheis. It is notable that Loch Nevis is a long way from Ben Nevis, and must have acquired its name independently: another argument for the water-origin.

New Abbey (Dumfries & Galloway) Named after the Cistercian abbey founded in 1273 by Devorguilla Balliol, buried here with the heart of her husband. Hence the romantic name 'Sweetheart Abbey'.

Newbattle (Midlothian) 'New building'. *Neowe* (Old English) 'new', *botl* 'house', 'dwelling'; the name dates from the foundation of the abbey in 1140. Noted as Niwe Bothla, 1141.

Newburgh (Fife; Aberdeenshire) 'New town'. *Neowe* (Old English) 'new', *burh* 'town'. The Fife town was granted a charter by Alexander III in 1266; the Aberdeenshire village dates back to 1261.

Newcastleton (Borders) Established in 1793 by the third duke of Buccleuch, relocating the former Castleton, noted as Cassiltoun in 1275.

New Deer (Aberdeenshire) *See* Deer for derivation of the name. The village dates from 1805, when it was established by James Ferguson of Pitfour.

New Galloway (Dumfries & Galloway) First recorded as the New Town of Galloway in 1682, its charter was granted by King Charles I in 1629 to Sir John Gordon.

Newhaven (Edinburgh) 'New harbour', founded by King James IV in 1510.

Newington (Edinburgh) 'New place'. The name is first found around 1720; its form might be expected to be Newton. The *'ing'* element may have been modelled on older placenames in the vicinity.

New Pitsligo (Aberdeenshire) *See* Pitsligo for derivation of the name. This location dates from 1780, when it was founded by Sir William Forbes of Pitsligo.

Newport-on-Tay (Fife) Established 1713, first as Newport Dundee, as a ferry port.

New Scone (Perth & Kinross) Created in 1805, when the old village was demolished in order for the earl of Mansfield's park to be extended. See Scone.

Newtongrange (Midlothian) So called in contradistinction to the older grange of the nearby Newbattle Abbey, Prestongrange near Prestonpans.

Newtonmore (Highland) 'New town on the moor'. A name dating from the mid-18th century.

Newton Stewart (Dumfries & Galloway) Established in 1671 by William Stewart, son of the earl of Galloway, who obtained a burgh charter from King Charles II.

Niddrie (Edinburgh) 'New house', or 'new farm'. *Newydd* (Brittonic) 'new', *tref* 'house', 'farmstead'. Found as Nodref, c. 1249. Niddry has the same derivation.

Nigg (Highland; Aberdeen) 'On the bay'. *An uig* (Gaelic adaptation of the Old Norse *vik*) 'bay'. The Ross-shire village has given the name back to Nigg Bay, as with the other Nigg, south of Aberdeen, noted c. 1250 as Nig.

Nith (Dumfries & Galloway) 'Glistening stream', deriving from Brittonic *Nedd* (cognate with Nethy) meaning 'glistening'. The Novios river shown in Ptolemy's map c. 150 is in the approximate position of the Nith, but it seems unlikely that the names are related. Noted as Nidd by Bede (731).

Nitshill (Glasgow) 'Nut hill.' *Nit* (Scots) 'nut.'

Noltland (Orkney) 'Cattle land'. *Nauta* (Old Norse) 'cattle', giving Scots *nolt* or *nowt*.

North Berwick (East Lothian) North 'barley farmstead'. *Bere* (Old English) 'barley', *wic* 'farmstead'. Recorded as Northberwyk in 1250. See Berwick-upon-Tweed.

Noss (Shetland; Highland) 'The nose'. *Nos* (Old Norse) 'nose'. Noss Head in Caithness has the same derivation.

Novar (Highland) Possibly 'Giant's house'. The Gaelic name is Taigh an Fhuamhair: *taigh* 'house'; *an* 'of'; *fhuamhair* 'giant's', or 'champion's'. It is noted on Blaeu's map, 1654, as Tenuer.

O

Oakley (Fife) 'Oak-field', from oak and *ley* (Middle English and Scots) 'meadow'.

Oban (Argyll & Bute) 'Little bay'. *Ob* (Gaelic) 'bay' *-an* (Gaelic diminutive suffix) 'little'. Its full Gaelic name is An t-Oban Latharnach, 'the little bay of Lorn'.

Ochil Hills (Clackmannanshire, Perth & Kinross) 'High hills'. *Ocel* (Brittonic) 'high'. The earliest references are to Cind Ochil in 700, Sliab Nochel 850, Oychellis 1461.

Ochiltree (East Ayrshire; West Lothian) 'High house'. *Ocel* (Brittonic) 'high', *tref* 'house', 'homestead'. Recorded as Okeltre, c. 1200.

Ochtertyre (Perth & Kinross) Scotticised form of Auchtertyre, 'upper land', from *uachdar* (Gaelic) 'upper', *tiridh* 'land'.

Oich, River and Loch (Highland) 'Stream-place'. *Abha* (Gaelic) 'stream', *ach* 'place'. Found in the present form 1769. The loch may have originally been *Loch Abha* (as in Awe). *See* Awe, Avoch.

Old Kilpatrick (West Dunbartonshire) 'Old place of St Patrick's church'. *Cill* (Gaelic) 'church'; *Padraig* (Irish Gaelic personal name). Noted as Kylpatrick, 1233. The village was formerly known simply as Kilpatrick, until the parish was split in 1649, since when it was prefixed first by 'West', and latterly by 'Old'.

Oldmeldrum (Aberdeenshire). *See* Meldrum. The 'Old' was added later when the former parish was divided.

Omoa (North Lanarkshire) Named by a local landowner, Captain John Dalrymple, who had fought in the capture of Omoa, in Honduras, in 1779.

Oransay (Argyll & Bute) 'Oran's island', from Odhran (Irish Gaelic personal name), and Old Norse *ey*, 'island'. Noted as Ornansay, 1549. There are numerous islands bearing forms of this name. Oronsay is also 'Oran's isle'.

Orbliston (Moray) The first part has been tentatively identified with *iorbull* (Gaelic) 'peaceful'; the form suggests a personal name with -*ton* (Scots) 'town, place'; but no document has yet been found to back this up.

Orchy, River and **Glen** (Argyll & Bute) An Old Gaelic compound name, *Urcháidh* 'woody stream place'. It has been seen as combining Gaelic *ar* 'on', 'near', with Old Gaelic elements *cet* indicating 'wood', and -*ia* 'stream'.

Orkney Possibly 'the Boar tribe's islands'. *Orc* (Celtic root noted in Latin texts of 320 BCE) 'boar', 'pig'; noted in Old Irish as Insi-orc, 'islands of the pigs' (*uirc*). The name was recorded around 30 BCE by Strabo, and by the Romans in the first century CE as Orcades. This latter name is still occasionally used in a literary context. Ptolemy, c. 150, has Orkades; the *Pictish Chronicle*, c. 970, has Orkaneya. It was assimilated into Old Norse as *Orkneyjar*, with the meaning of 'seal islands'. The Gaelic name is Arcaibh.

Ormiston (East Lothian) 'Orm's farm'. *Ormr* (Old Norse proper name), *tūn* (Old English) 'farmstead'. Found in this form in 1293.

Orphir (Orkney) 'Tidal island'. Old Norse *orfiris* composed of *or* 'out of', and *fjara* 'foreshore'.

Orrin, River and **Glen** (Highland) A suggested origin is 'river of the chapel', from the church of Urray, *oifreann* (Gaelic), 'chapel', 'offering place', at the point of confluence with the Conon, recorded in 1440 as Inverafferayn . But it is likely to be much older, deriving perhaps from the pre-Celtic root-form **ora*, indicating 'flowing'. See Lochore.

Orton (Moray) 'Edge of the hill'. *Oir* (Gaelic) 'edge', *dhùin* ' of the hill'. Noted as Urtene, 1542.

Otter Ferry (Argyll & Bute) 'Ferry of the reef'. *Oitir* (Gaelic) 'reef'. Noted as Ottyr, 1490.

Oxgangs (Edinburgh; other districts) Scots name for an area of land that could be ploughed by an ox in a day.

Oxnam (Borders) 'Ox farm'. *Oxenaham* (Old English) 'village of the oxen', noted c. 1150 as Oxanaham.

Oykel, River and **Strath** (Highland) Perhaps as Ochil, from a lost Pictish word related to Brittonic *ocel* 'high'. It is no longer considered the battle-site *Ekkjalsbakki* (Old Norse) 'Ekkjal's bank', in the *Orkneyinga Saga*. It was noted as Okel, 1365.

P

Pabay (Western Isles) 'Priest's island'. *Papa* (Old Norse) 'priest', *ey* 'island'. Also Pabbay.

Padanaram (Angus) A Biblical name given in the seventeenth century. *See* Joppa.

Paisley (Renfrewshire) 'Pasture slope', from *pasgell* (Brittonic) 'pasture', and *llethr* 'slope', cognate with Gaelic *leitir*, has been suggested as the most likely derivation. Recorded as Passeleth 1157, Paislay 1508.

Panmure (Angus) 'Big hollow'. *Pant* (Brittonic–Pictish) 'hollow, dene', *mawr* 'big'. Noted as Pannemore, 1286.

Papa Stour (Shetland) 'Great priest island'. *Papa* (Old Norse) 'priest', *ey* 'island', *storr* 'great'. 'Storr' distinguishes it from the other Old Norse 'priest islands'; noted in 1229 as Papey Stora.

Partick (Glasgow) 'Bushy place'. *Perth* (Pictish) 'thicket'. Early records show Perdyec 1136, Pertheck 1158, Perthik 1362.

Patna (East Ayrshire) The name of an Indian city, borrowed 1810 by the local landowner, William Faulkner.

Peebles (Borders) Place of 'sheilings'. Perhaps from *pebyll* (Brittonic) 'sheilings', 'tents'. Noted as Pobles, c. 1124, Pebles c. 1126.

Pencaitland (East Lothian) Possibly 'head of the wood enclosure'. *Pen* (Brittonic) 'head', 'top of', *coet* 'wood', and *lann* 'field', 'enclosure'. Found as Pencatlet, c. 1150.

Penicuik (Midlothian) 'Hill of the cuckoo'. *Pen* (Brittonic) 'head, hill', *y* 'the', *cog* 'cuckoo'. Recorded as Penicok in 1250. The Scots word for cuckoo, 'gowk', is found at Gowkley Moss, a mile north of the town.

Pennan (Aberdeenshire) 'Headland water'. *Pen* (Brittonic–Pictish) 'headland', 'hill', *an* 'water, stream'. A *pen-* name is most unusual so far north; if Pictish, it is an almost unique survival.

Pennyghael (Argyll & Bute) Taken to mean 'Penny-rental land of the Gael'. *Peighinn* (Gaelic) 'penny' (rental). Some land in the area must have been held by non-Gaels.

Penpont (Dumfries & Galloway) 'Head, or end, of the bridge'. *Pen* (Brittonic) 'head', *pont* 'bridge'. The name suggests a bridge of great antiquity, quite possibly a Roman one.

Pentland Firth (Highland; Orkney) 'Firth of Pictland'. *Pettr* (Old Norse) 'Picts', *land* 'territory, *fjordr* 'sea inlet' or 'passage'. Noted as Pettlandsfjordr, c. 1100. The name really ought to be Pictland Firth.

Pentland Hills (Midlothian) 'Hill-land'. *Pen* (Brittonic) 'hill'; *land* (Old English) 'tract of land'. Recorded as Pendant, c. 1150. The name bears no relation to that of the Pentland Firth.

Perth (Perth & Kinross) 'Bushy place.' *Perth* (Pictish) 'bush', 'thicket'. Recorded as Pert in 1128. For several hundred years known as St Johnstoun, or St John's Toun of Perth, after the building of St John's Kirk in the 12th century.

Peterculter (Aberdeen) 'Corner land of St Peter'. *Cuil* (Gaelic) 'corner', *tir* 'land'. The 'Peter', referring to the dedicatee of its church, was added later to distinguish it from the nearby village of Maryculter, hence its English form.

Peterhead (Aberdeenshire) 'St Peter's headland'. Founded in 1593, taking its name from St Peter's Kirk built here in 1132. Noted in 1544 as Petyrheid, 1595 as Peterpolle, from Scots *poll* or *pow* 'head'.

Philiphaugh (Borders) 'Shut-in valley'. *Ful* (Old English) 'closed', *hop* 'enclosed valley'. Recorded in the thirteenth century as Fulhope. Scots *haugh*, indicating low ground by a stream, is a later addition.

Pierowall (Orkney) 'Little bay'. *Piril* (Old Norse) 'small', and *vágr* 'bay'.

Pinkie (East Lothian) 'Cé's height'. *Pen* (Brittonic) 'hill'. *Cé* (Brittonic personal name). Noted as Penke, c. 1260.

Pinmore (South Ayrshire) 'Big penny-land'. *Peighinn* (Gaelic) 'penny, denomination of land equal to a penny rental' and *mór* 'great'. Pinwherry is 'Pennyland of the copse', *an fhoithre* 'of the copse'.

Pit- A prefix found almost entirely on the eastern side of the country from Fife to south-east Sutherland, within Pictish-ruled territory until AD 843. It comes from Pictish *pett*, 'portion of land' (earliest record in *The Book of Deer*, c. 1150). Not found in other Celtic placenames, it has been taken as a sign that Pictish was a separate (though related) language to Brittonic. The suffix in most Pit-names is Gaelic, suggesting that the *pit-* form was maintained since it had a meaning that was not conveyed by any Gaelic word. But in many instances an earlier Pit- may have been replaced by Gaelic *Baile*.

Pitagowan (Perth & Kinross) 'The smith's portion'. *Pett* (Pictish) 'portion', 'piece of land'; *ghobhainn* (Gaelic) 'smith's.'

Pitcairn (Fife) 'Portion of the cairn'. *Pett* (Pictish) 'portion', 'piece of land', *càrn* (Gaelic) 'cairn'. Found as Peticarne, 1247.

Pitcaple (Aberdeenshire) 'Horse share'. *Pett* (Pictish) 'portion', 'piece of land'; *capull* Gaelic 'horse'.

Pitfour (Angus; Highland) 'Pasture share'. *Pett* (Pictish) 'portion', 'piece of land'; *phúir* (Scottish Gaelic) 'pasture'. The Black Isle name is found as Pethfouyr, c. 1340.

Pitkeathly (Perth & Kinross) 'Cathalan's land'. *Pett* (Pictish) 'portion', 'piece of land'; *Cathalan* (Irish Gaelic personal name). Found as Pethkathilin, c. 1225.

Pitkerrow (Perth & Kinross) 'Fourth part of land'. *Pett* (Pictish) 'portion', 'piece of land'; *ceathramh* (Gaelic) 'fourth part'.

Pitlochry (Perth & Kinross) 'Piece of land by or with the stones'. *Pett* (Pictish) 'portion', 'piece of land'; *cloichreach* (Gaelic) 'stones'. The 'stones' were probably stepping stones across the River Tummel.

Pitmaduthy (Highland) 'MacDuff's portion'. *Pett* (Pictish) 'portion', 'piece of land'; *mhic Dhuibh* (Gaelic) 'MacDuff's'.

Pitscottie (Fife) 'Portion of flowers'. *Pett* (Pictish) 'portion', 'piece of land'; *sgothaich* (Gaelic) 'flowery'. Found as Petscoty, 1358.

Pitsligo (Aberdeenshire) 'Shelly portion'. *Pett* (Pictish) 'portion', piece of land'; *sligeach* (Gaelic) 'shelly'. Found as Petsligach, 1426.

Pittendreich (Fife; West Lothian; Moray) 'Portion of the good aspect'. *Pett* (Pictish) 'portion', 'piece of land'; *dreach* (Gaelic) 'aspect, beauty'. This is the most common Pit- name. Variant forms include Pittendrigh, Pendreich. The Lothian place is noted as Petyndreih, 1140.

Pittenweem (Fife) 'Place of the cave'. *Pett* (Pictish) 'portion', piece of land'; *na* (Gaelic) 'of the', *h-uamha* (Gaelic) 'cave'. Recorded in 1150 as Petnaweem.

Pittodrie (Aberdeen) 'Portion by the woodland'. *Pett* (Pictish) 'portion', 'piece of land'; *fhodraidh* (Gaelic) 'by the wood'.

Plockton (Highland) 'Town of the block'. *Ploc* (Gaelic) 'block', with *-ton* (Scots) 'place'. Its Gaelic name is Am Ploc.

Pluscarden (Moray) 'Big house of the thickets'. *Plas* (Pictish-Brittonic) 'hall'; *cardden* (Pictish) 'thicket'. Recorded in 1124 as Ploschardin.

Pollokshaws (Glasgow) 'Pool by the thicket'. *Poll* (Brittonic) 'pool', *-oc* (diminutive suffix) 'little'; *sceaga* (Old English) 'wood'.

Polmont (Falkirk) 'Pool-hill'. *Poll* (Gaelic) 'pool', 'hollow', *monadh* 'hill'. Noted as Polmunth, 1319.

Pomona (Orkney) An alternative name for the Orkney Mainland, first found in Fordun's *Chronicle*, c. 1380, and current until the nineteenth century.

Poolewe (Highland) 'Pool of the (river) Ewe'. *Poll* (Gaelic) 'pool', 'hollow', *iu* river-name perhaps related to 'yew'. *See* Kinlochewe.

Port Bannatyne (Argyll & Bute) Named for the Bannatyne family who established their seat at the nearby Kames Castle in the thirteenth century.

Port Charlotte (Argyll & Bute) Named after Lady Charlotte, mother of the Gaelic scholar W. F. Campbell of Islay, who founded the village in 1828.

Port Ellen (Argyll & Bute) Named after Lady Ellenor, wife of W. F. Campbell, who founded the village in 1821.

Port Glasgow (Inverclyde) Founded by Glasgow Town Council in the 1660s in order to provide a deep-water port for the city's developing Atlantic trade.

Port Seton (East Lothian) 'Seton's harbour'. The Setons were landowners who developed the harbour from the 15th century onwards.

Portavadie (Argyll & Bute) 'Beaching place of the dogs, or foxes', from Gaelic *port*, 'harbour, beaching place', and *mhadhaidh*, 'of dogs'.

Portessie (Moray) 'Port of the waterfall'. *Port* (Gaelic) 'harbour', 'beaching place', *easach* 'waterfall', 'tumbling stream'.

Portgower (Highland) Clearance village named from the duke of Sutherland's family name, Leveson-Gower.

Portincaple (Argyll & Bute) 'Port of the horse'. *Port* Gaelic) 'harbour', 'beaching-place', *nan* 'of the', *chapuill* 'horse'. Found as Portinkebillis c. 1350; also as Portinchapil, which may indicate a false etymology based on 'chapel'.

Portknockie (Moray) 'Harbour by the little hill'. *Port* (Gaelic) 'harbour', *cnoc* 'rounded hill', *-ie* (colloquial diminutive) 'little'. This fishing port was founded in 1677.

Portlethen (Aberdeenshire) 'Port of the slope'. *Port* (Gaelic) 'harbour', 'beaching place', *leathan* 'slope'.

Portmahomack (Highland) 'Haven of (saint) Colman'. *Port* (Gaelic) 'harbour', 'beaching place'; *mo* 'my', *Cholmáig* (Irish Gaelic) 'Colman's' or 'Colum's'. Noted as Portmachalmok, 1678.

Portnahaven (Argyll & Bute) 'Harbour of the river'. *Port* (Gaelic) 'harbour', 'beaching place', *na* 'of', *abhainn* 'river'.

Portnalong (Western Isles) 'Port of the ships'. *Port* (Gaelic) 'harbour', 'beaching place', *nan* 'of', *long* 'ship'.

Portobello (Edinburgh) From a house built here by a sailor who had seen action at the battle of Puerto Bello in Panama, 1739.

Portpatrick (Dumfries & Galloway) 'Harbour of St Patrick'. *Port* (Gaelic) 'harbour'; *Padraig* (Irish Gaelic personal name, from Latin *Patricius*, 'nobly-born') to whom a chapel was dedicated here.

Portree (Highland) 'Harbour of the slope'. *Port* (Gaelic) 'harbour', *righeadh* 'of the slope'. The second element is not thought to derive from *rígh* 'king'.

Portsoy (Aberdeenshire) 'Hay harbour', from Gaelic *port* 'harbour', and *saoidh* (feminine) 'hay' is perhaps more likely than from *saoidh* (masculine) 'righteous one'.

Prestonpans (East Lothian) 'Priests' village by the salt-pans'. *Preost* (Old English) 'priest', *tūn* 'village'; *-pans* (Scots) 'salt-pans'. Recorded as Saltprestoun 1587.

Prestwick (South Ayrshire) 'Priests' farm'. *Preost* (Old English) 'priest', *wic* 'farm'. Noted as Prestwic, 1170.

Q

Quanterness (Orkney) 'Bishop's point'. *Kantari* (Old Norse) 'bishop' (from 'Canterbury'), *nes* 'cape', 'point'.

Quarff (Shetland) 'Shelter'. *Hvarf* (Old Norse) can mean 'shelter place' as well as 'turning' (see Wrath).

Queensferry, North and **South** (Fife and Edinburgh) Named for Queen Margaret, wife of King Malcolm III, who established a ferry here. Noted as Queneferie, c. 1295.

Quinag (Highland) 'Milk pail', from Gaelic *cuinneag*, 'milk pail'. *Caoin* (Gaelic) 'fair', 'beautiful', has also been suggested.

Quiraing (Highland) 'Crooked enclosure'. *Kví* (Old Norse) 'cattle-fold', *rang* 'crooked'.

Quoich, River, Glen and **Loch** (Highland) Probably 'of the hole(s)'. *Cuaich* (Gaelic) 'of the hollows', 'pot-holes'. Originally the river name.

R

Raasay (Highland) 'Roe-deer ridge island'. *Rar* (Old Norse) 'roe deer', *ass* 'ridge', *ey* 'island'. Noted as Raasa 1263, Rairsay 1526.

Rackwick (Orkney) 'Sea-wrack bay'. *Reka* (Old Norse) 'sea wrack, sea-weed', *vik* 'bay'. Found as Rekavik in the *Orkneyinga Saga*, c. 1225.

Rafford (Moray) 'High fort'. *Rath* (Gaelic) 'ring fort', *àird* 'high'. Noted as Raffart, c. 1700.

Raith (Fife) Probably 'Ring-fort'. *Rath* (Gaelic) 'ring-fort'. Noted as Rathe, c. 1320.

Rankeilour (Fife) 'Portion by the clay stream'. *Rann* (Gaelic) 'part', 'portion'; *cil* 'red clay', *dobhar* 'stream'. Noted as Rankeloch, 1293. Also found as Rankeillor.

Rannoch, Moor and Loch (Highland; Perth & Kinross; Argyll & Bute) 'Bracken'. *Raineach* (Gaelic) 'bracken', 'fern'.

Ranza, River and **Loch** (North Ayrshire) 'Rowan tree river'. *Reynis* (Old Norse) 'rowan', *áa* 'river'. Noted as Lockransay, 1433.

Rattray (Perth & Kinross; Aberdeenshire) 'Homestead of the ring-fort'. *Rath* (Gaelic) 'ring-fort'; *tref* (Brittonic–Pictish) 'homestead'. Found as Rotrefe, 1291. The same name occurs in Rattray Head, with Old Rattray inland.

Reay (Highland) 'Ring-fort'. *Rath* (Gaelic) 'ring-shaped stone fort'. Noted as Ra, c. 1230.

Renfrew (Renfrewshire) 'Point of the current'. *Rhyn* (Brittonic) 'point', *frwd* 'current'. Records show Renifry 1128, Reinfrew 1158, Renfrew 1160.

Renton (West Dunbartonshire; Borders) The Dunbarton name was given in 1782 by its founder, Jean Smollett, after her daughter-in-law, Cecilia Renton. The Borders name is from *Regna* or *Regenhild* (Old English personal name), *ing* indicating ownership, *tūn* 'farm', recorded as Regninton from the 11th century.

Restenneth (Angus) 'Moor of fire'. *Ros* (Brittonic–Pictish) 'moor, *tened* 'of fire'. Recorded as Rostinoth, c. 1150.

Reston (Borders) 'Rhys's place'. *Rhys* (Brittonic personal name); *tūn* (Old English) 'settlement', 'farm'.

Rhu (Argyll & Bute) 'Cape', 'headland'. *Rudha* (Gaelic) 'cape'.

Rhynie (Aberdeenshire; Highland) In Aberdeenshire found as Ryny, c. 1230, as 'division or portion of land', from Gaelic *roinnean*, a diminutive of *roinn* 'share', 'portion of land'. In Ross-shire it is from Gaelic *rathan*, 'small fort'. Found as Rathne, 1529.

Riccarton (Borders; East Ayrshire) 'Richard's place'. *Riccart* (Old English personal name), *tūn* 'settlement, farm'. The Kilmarnock site is named after Sir Richard Wallace, an uncle of William Wallace.

Rinnes, Ben (Moray) Perhaps 'promontory hill'. *Beinn* (Gaelic) 'mountain', and *rinn* 'promontory', 'point'; alternatively 'hill of the shares', from *roinn* 'division'. In Gaelic it is Beinn Rinneis.

Rinns of Galloway (Dumfries & Galloway) Probably 'promontories'. *Rinn* (Gaelic) 'point', 'promontory'. Two other Gaelic words are possible sources, either *rann*, or *roinn;* both can mean 'part', 'portion'. Noted 1460 as le Rynnys. Also spelt Rhinns, Rhynns. The Rinns of Islay share these possible derivations.

Risk (Dumfries & Galloway) 'Marsh', 'bog'. *Riasg* (Gaelic) 'morass'. Related forms include Reisk, Risko, Ruskich, Ruskie (with -*ach* suffix denoting place).

Robroyston (Glasgow) 'Robert's place'. Noted as Roberstoun in the 16th century.

Rockall (Western Isles) Perhaps 'Bare island in the stormy sea'. *Rok* (Old Norse) 'stormy sea', *kollr* 'bald head'.

Rodil (Western Isles) Perhaps 'roe deer valley'. *Rá* (Old
Norse) 'roe', *dalr* 'valley'. Noted as Roadilla, 1580.
Raadil might be expected; *see* Raasay.

Rogart (Highland) 'Red enclosure'. *Raudr* (Old Norse)
'red', 'reddish', *gardr* 'enclosure', 'garth'. Noted as
Rothegorth, c. 1230.

Rohallion (Perth & Kinross) 'Ring-fort of the Caledonians'.
Ràth (Gaelic) 'ring-fort', *chaileainn* 'of the Caledonians'.

Romanno (Borders) 'Fort of the monk'. *Ràth* (Gaelic)
'ring-fort', *manaich* 'of the monk'. Noted as Rothmanaic,
c. 1160.

Rona (Highland; Western Isles) 'Rough rocky island'. *Hraun*
(Old Norse) 'rough', 'rocky', *ey* 'island'.

Ronaldsay (Orkney) North Ronaldsay is 'Ninian's isle'.
Ringan (Old Norse form of Ninian), *ey* 'island'.
Recorded as Rinarsey in *Orkneyinga Saga*, c. 1225. South
Ronaldsay is 'Rognvaldr's isle'. *Rognvaldr* (Old Norse
personal name), *ey* 'island', recorded as Rognvalsey in
the Norse sagas.

Rosehearty (Aberdeenshire) 'The point, or wood, of
Abhartach' (Old Gaelic personal name), prefixed by *ros*
(Gaelic) 'point', 'wood'. Found as Rossawarty, 1508.

Rosemarkie (Highland) 'Promontory of the horse'. *Ros*
(Gaelic) 'promontory', *marc* 'horse'. Just inland is
Drummarkie, 'ridge of the horse'.

Roslin (Midlothian) 'Holly moor'. *Ros* (Brittonic) 'moor',
celyn 'holly'. Recorded as Roskelyn, c. 1240.

Rosneath (Argyll & Bute) 'Point of the sanctuary'. *Ros*
(Gaelic) 'promontory', *neimhidh* 'of the holy place'.
Noted simply as Neveth, c. 1199, Rosneth 1225.
See Navitie.

Ross (Highland; Argyll & Bute; Dumfries & Galloway)
Early forms occur in Brittonic and Gaelic. Brittonic *ros* or
rhos means 'promontory' or 'moor'; Gaelic *ros* can mean
both 'promontory' and 'wood'. The name of the former
county is generally taken to mean 'promontory' from its
long eastern peninsulas. The locations of the Ross of Mull
and the Ross names on the western tip of Kirkcudbright
Bay also suggest the 'promontory' meaning.

Rosyth (Fife) 'Headland of Fife'. Gaelic *ros* 'promontory', with area-name *Fiobha* from Pictish *Fib* (personal name). Noted as Rossyth, c. 1130. See Fife.

Rothes (Moray) 'Ring fort'. *Rath* (Gaelic) 'ring fort'. Found as Rothes, 1238.

Rothesay (Argyll & Bute) 'Rotha's isle'. *Rotha* (Old Norse personal name), *ey* 'island'. Found as Rothersay, 1321, it originally referred to Rothesay Castle, which is still surrounded by a moat, and was extended to the town, whose Gaelic name is Baile Bhòid, 'town of Bute'.

Rothiemurchus (Highland) 'Muirgus's fort'. *Ràth* (Gaelic) 'ring fort'; *Muirgus* (Old Gaelic personal name). Recorded as Rathmorchus, 1226.

Rousay (Orkney) 'Hrolfr's island'. *Hrolfr* (Old Norse personal name), *ey* 'island'. Found in 1260 as Hrolfsey.

Rowardennan (Stirling) 'Adamnan's high promontory'. *Rudha* (Gaelic) 'cape', 'promontory', *àird* 'height', *Adhamhnáin* personal name) 'of Adamnan', St Columba's biographer, d.704.

Roxburgh (Borders) 'Hroc's fortified dwelling'. *Hroc* (Old English personal name), *burh* 'fortified dwelling'. Recorded as Rokisburc in 1127.

Roy, River and **Glen** (Highland) 'Red River'. *Ruaidh* (Gaelic) 'red'.

Rubislaw (Aberdeen) 'Rubie's Hill'. *Rubie* (Scots diminutive of Reuben), *law* 'hill'.

Ruchil (Perth & Kinross; Glasgow) The Tayside Ruchil is 'red flood'; *ruadh* (Gaelic) 'red'; *thuil* 'flooding stream'. The Glasgow district is 'red wood', with the second part being *choille* (Gaelic) 'wood'. It is sometimes found as Ruchill.

Rum (Highand) Perhaps 'spacious (island)'. *Rùm* (Gaelic) 'room', 'space'. But a lost pre-Celtic origin has also been suggested. The alternative spelling, 'Rhum' was attempted by an early 20th-century owner. Recorded as Ruim in 677.

Rutherglen (South Lanarkshire) 'Red valley'. *Ruadh* (Gaelic) 'red', *gleann* 'glen, valley'. The form Ruthirglen is found in 1153.

Ruthven (Highland; Perth & Kinross) 'Red river', from Gaelic *ruadh* 'red' and *abhainn* 'river'.

Ruthwell (Dumfries & Galloway) 'Spring of the Holy Cross', from *róde* (Old English) 'cross', 'rood'; and *waelle* 'spring'.

Ryan, Loch (Dumfries & Galloway) 'Of the chief', is a possible derivation for the name, noted on Ptolemy's map of c. CE 150 as Rerigonios Kolpos (Greek 'gulf'); with Brittonic *rigon*, 'chief', cognate with Welsh *rhion*.

S

Saddell (Argyll & Bute) Noted as Sagadul, 1203, perhaps from *sag* (Old Norse) 'saw' and *dalr* 'valley', 'dale': a place for sawing timber.

St Abb's Head (Borders) Named after Aebba, the first prioress of Coldingham, and sister of the seventh-century King Oswald of Northumbria.

St Andrews (Fife) Scots 'Sanct Androis'. Prior to its 9th century elevation as a Christian cult centre, it was Mucros (Gaelic) 'wood, or point, of the pigs'. A local hill was Ceannrighmonaigh, 'top of the king's mound' *(Annals of Tighernach)*. The *righ* element has also been read simply as 'slope', The prefix changed to Cillrighmonaidh (Gaelic *cill*) after founding of the church, and the name became Chilrimunt, c. 1139, then Kilrymont, displaced by St Andrews as town name in the 14th century. In Gaelic it is Baile Reuil, 'town of St Rule'.

St Boswells (Borders) Named after St Boisil, the 7th-century abbot of Melrose and friend of St Cuthbert. The -wells part is from Norman French *-vil* or *ville*, 'town'.

St Cyrus (Angus) 'Church dedicated to St Cyricus'. Another early form is Ecclesgreig, from a conjectured Brittonic–Pictish *egles*, 'church', *Girig* (Pictish personal name): 'Girig's church'. Found as Eglesgreig, 1243 The church was dedicated by Girig, a 9th-century Pictish king, to St Cyricus.

St Fillans (Perth & Kinross) 'Place dedicated to St Fillan'. *Faolán* (Irish Gaelic personal name) literally 'Little Wolf', 8th century Irish missionary. The village lies in Strathfillan, recorded as Strathfulane, 1317, 'Fillan's strath'.

St Fort (Fife) 'Sand ford'. Noted as Sandfurde in 1449. In the 17th century it is recorded as Santford, and this appears to have occasioned the present misleading form.

St Kilda (Western Isles) Not named after a St Kilda – no saint of this name is known: it is Old Norse *skildir*, 'shields'. Originally referring to islands closer inshore, it was transferred to the remoter archipelago. Also suggested is *kelda* (Old Norse) 'well', as the landing place on Hirta, the main island, is in Gaelic Tobar Childa, which in the post-Norse era may have been misunderstood as 'Kilda's well', though it simply means 'well' in both languages. But it does not explain the Sk- element. The Gaelic name of the island group was and is *Hirt*. See Hirta.

St Margaret's Hope (Orkney) 'St Margaret's Bay'. *Hop* (Old Norse) 'bay', 'hollow place'. Named for Margaret, consort of Malcolm III.

St Monans (Fife) Its church is dedicated to St Monan, the 6th-century bishop of Clonfert in Ireland. Noted as Sanct Monanis, 1565, the name was later spelt as St Monance, until the twentieth century.

St Ninian's Isle (Shetland) The missionary activities of Ninian and his followers along the east coast have left traces in local names all the way to this most northerly one. Perhaps because of the island's small size, its name escaped the alteration undergone by North Ronaldsay.

St Rollox (Glasgow) 'St Roche'. A chapel to St Roche was set up here in 1502.

St Vigeans (Angus) Perhaps a form of St Féchín, an Irish saint of the mid-7th century.

Salen (Argyll & Bute) 'Inlet'. *An* (Gaelic) 'the', *sailein* 'little inlet of the sea'.

Saline (Fife) 'Salt pit'. In 1613 the name is recorded as Sawling.

Saltcoats (North Ayrshire) Place of the 'salt huts'. *Salt* refers to the process of saline extraction (see Prestonpans); *cots* (Scots) 'cottages', 'huts'. Noted as Saltcotes, 1548.

Sanday (Orkney) 'Sand island', from *sand* (Old Norse) 'sand', and *ey* 'island'. Found in this form 1369.

Sanquhar (Dumfries & Galloway) 'Old fort'. *Sean* (Gaelic) 'old'; *caer* (Brittonic) 'fort'. Recorded as Sanchar c. 1150.

Sauchieburn (West Lothian) 'Burn by the willows'. *Saileach* (Gaelic) 'willow'; *burn* (Scots) 'stream'.

Saughton (Edinburgh) 'Place by the willows'. *Saileach* (Gaelic) 'willow'; *tūn* (Old English) 'settlement', giving Sots -*ton*.

Scalloway (Shetland) 'Bay of the huts'. *Skali* (Old Norse) 'huts', *vágr* 'bay'. Temporary huts were erected here by those attending the assemblies held at nearby Tingwall.

Scalpay (Western Isles, Orkney) 'Ship isle'. *Skálp* (Old Norse) 'boat', 'ship', *ey* 'island'. Scalpay in Orkney is 'ship isthmus': the suffix being from *eidh* (Old Norse) 'isthmus'.

Scapa Flow (Orkney) Probably 'Sea-flood bay of the boat isthmus'. *Skálp* (Old Norse) 'boat', *eidh* 'isthmus', *floa* 'flood'.

Scarba (Argyll & Bute) 'Cormorant island'. *Skarfr* (Old Norse) 'cormorant', *ey* 'island'. Skarfskerry has the same derivation, with *skjaer* 'rock'.

Scavaig, River and **Loch** (Highland) The derivation of the first part is unclear, perhaps linked to an Old Norse root *ska*, 'scrape'. The suffix is *vik* (Old Norse) 'bay'. It may be an old river-name with -*vaig* back-formed on to it.

Schiehallion (Perth & Kinross) Perhaps 'maiden's pap', from *sine* (Gaelic) 'breast', and *chailean* 'girl's'. But it has also been related to *sithean* (Gaelic) which can mean 'fairy hill', with *chailleainn* 'of the Caledonians' (see also Ben Ledi). Noted as Schachalzean, 1642.

Sciennes (Edinburgh) A corrupt form of 'Siena'; referring to the former monastery of St Catherine of Siena. Noted in the 16th century as Shenis.

Scone (Perth & Kinross) 'Mound'. *Sgonn* (Gaelic) 'mound', 'lump'. The reference is to the Mote Hill, an ancient ritual site of Scottish kings. Recorded as Sgoinde in 1020.

Scotia 'Scotland'. A latinised form of 'land of the Scots', originally applicable to Ireland, then to medieval Scotland, and nowadays used in a poetic, antiquarian or fanciful sense for Scotland.

Scotland 'Land of the Scots'. The original Scots *(Scoti* in Latin) were 5th-6th century Gaelic-speaking immigrants. 'Scot' was legendarily supposed to derive from an ancestress Scota, daughter of an Egyptian pharaoh. 'Scotland', with its Anglian suffix *-land,* was probably coined by the Anglian-speaking population of Northumbria and Lothian. In Gaelic the name was and remains Alba.

Scotsburn (Highland) Perhaps a Scots rendering of *Allt* (Gaelic) 'stream', *nan* 'of', *Albannaich* 'the Scots'; itself a Gaelic translation of a Pictish name, indicating the site of a land-division between Picts and Gaelic-speaking Scots.

Scotscalder (Highland) 'Calder of the Scots' (see Calder). Nearby there was a Norn Calder, respectively noted in 1538 as Scottiscaldar, Nornecaldar; originally demarcating land held by those of Scots and Norse extraction in this once Norse region.

Scotstarvit (Fife) This hybrid name indicates 'Scot's bull place'. *Tarbh* (Gaelic) 'bull', *ait* suffix indicating place. The *Scot-* prefix may come from the land-owner's name. South of Cupar there is a spread of 'bull' names on each side of Tarvit Hill.

Scourie (Highland) 'Place of the wood'. *Skógr* (Old Norse) 'wood'. The Gaelic name is *Sgobhairidh.*

Scrabster (Highland) 'Rocky farmstead'. *Skjaere* (Old Norse) 'rocks', *bolstadr* 'farmstead'. A personal name or nickname, Skara, has also been suggested for the prefix. In the *Orkneyinga Saga* (c. 1225) it is Ská-ra-bólstadr.

Seaforth, Loch (Western Isles) 'Loch of the salt-lagoon-firth'. This apparently tautological name is explained by the semi-landlocked part of the loch, *saer* (Old Norse) 'salt lake', with *fjordr* 'firth', 'fiord'. Gaelic 'loch' was added in the post-Norse era.

Selkirk (Borders) 'Church by the hall'. *Sele* (Old English) 'hall', 'manor house'; *cirice* 'church', becoming Scots 'kirk'. Recorded as Selechirche in 1124.

Shandon (Argyll & Bute) 'Old fort'. *Sean* (Gaelic) 'old', *dùn* 'fort'.

Shandwick (Highland) 'Sand bay'. *Sand* (Old Norse) 'sand', *vik* 'bay'.

Shanter (South Ayrshire) 'Old land', from *sean* (Gaelic) 'old', *tìr* 'land'; probably by comparison with 'new land' taken into cultivation.

Shapinsay (Orkney) 'Hjalpand's island'. *Hjalpand* (Old Norse personal name, *ey* 'island'. Noted in the *Orkneyinga Saga* (c. 1225) as Hjalpandisay.

Shawbost (Western Isles) 'sea-lake-farm'. *Sjá* (Old Norse) 'lake partly open to the sea', *bolstadr* 'farmstead'.

Shawfield (Glasgow) 'Wood field'. *Shaw* (Scots) 'wood' from *sceaga* (Old English) 'thicket', 'wood'.

Shee, Glen (Perth & Kinross) 'Fairy glen', or 'glen of peace'. *Gleann* (Gaelic) 'valley', 'glen'; *sìth* meaning 'fairy' or 'spirit', or 'peace'. The river in Glen Shee is the Blackwater, and perhaps this is another 'black goddess' stream *(see* Lochty).

Shetland Perhaps 'Hilt land'. The Old Norse name was 'Hjaltland', *Hjalt* 'hilt of a sword' or 'dagger', *land* 'land'. While the first *l* was dropped, the initial *Hj* was mutated to *Sh*. The Norse form is partly preserved in 'Sheltie', referring to the Shetland breed of pony. The oldest recorded Gaelic name is Innse Cat, 'islands of Cat'; latterly it is Sealtainn.

Shettleston (Glasgow) Originally noted in a Latin document of 1170 as 'the vill or villa of Seadna's daughter'. *Villa* (Latin 'house, place'*; nighine* (Gaelic) 'of a daughter', *Seadna* (Gaelic personal name). The form Inienchedin is found in 1186. By 1515 *villa* was translated into Scots *-ton*, 'township', as Schedilstoune.

Shiant Isles (Argyll & Bute) 'Charmed islands'. *Na-Eileanan* (Gaelic) 'the islands', *seunta* can mean 'under enchantment' or 'blessed'.

Shiel, River, Loch and **Glen** (Highland) Probably 'flowing water' from a Pictish word stemming from the continental Celtic root-form **sal* 'flowing'. Gaelic *seileach* 'willow' stems from the same source.

Shieldaig (Highland) 'Herring bay'. *Sild* (Old Norse) 'herring', 'sild'; *aig* (Gaelic form of Old Norse *vik)* 'bay'. Its Gaelic form is Sildeag.

Shin, River and **Loch** (Highland) From a pre-Celtic root-word *sinn*, indicative of flowing water, also the source of Irish 'Shannon'. Recorded in 1595 as Shyn.

Shiskine (North Ayrshire) 'Marshy place'. *Sescenn* (Gaelic) 'marsh'. Noted as Cescen, c. 1250.

Shotts (North Lanarkshire) Place of 'steep slopes'. *Sceots* (Old English) 'steep slopes'.

Sidlaw Hills (Perth & Kinross; Angus) 'Hills of seats' has been suggested, with the first element from *suidhe* (Gaelic) 'seat', often used in association with the name of a holy man, and Scots *law* 'hill'. 'Hills' is a linguistically superfluous later addition.

Skara Brae (Orkney) 'Bank by the shore'. *Skari* (Old Norse) 'shore'; *brae* (Scots) 'bank'.

Skelbo (Highland) 'Shelly farm'. *Skel* (Old Norse) 'shell', *bol* (abbreviated form of *bolstadr)* 'farm'. Noted as Scelbol, 1214, Scellebol c. 1300.

Skelmorlie (North Ayrshire) 'Scealdamèr's meadow'. *Scealdamer* (Old English personal name); *ley* (Scots) 'meadow'. Noted as Skelmorley, c. 1400.

Skene, Loch (Aberdeenshire; Dumfries & Galloway) Perhaps 'loch of hawthorns', from *loch* (Gaelic) 'lake', 'loch', and *sgeachan* 'of hawthorns'. The Aberdeenshire loch is known as Loch of Skene, found in this form 1318.

Skipness (Argyll & Bute) 'Ships' headland'. *Skipa* (Old Norse) 'ship', *nes* 'headland'. Recorded as Schepehinch, c. 1250, Skipnish 1260.

Skye (Highland) Perhaps 'Winged Isle'. *Sgiathach* (Gaelic) 'winged'. The reference appears to be to the shape of the island, with its many peninsulas. It is noted by Ptolemy, c. CE 150, as Ski or Skitis; and as *Scia* in Adamnan's *Life of St Columba* (c. 700). The familiar Gaelic name is *Eilean a'Cheo* 'isle of mist'.

Slamannan (Falkirk) 'Hill of Mannan'. Manau was the Brittonic name of the area at the head of the Firth of Forth, cognate with the (Isle of) Man. The prefix was probably Brittonic *mynnyd* 'hill', changed to Gaelic *sliabh*. The form in 1250 was Slethmanin.
See Clackmannan.

Slapin, Loch (Highland) The derivation may be from *slappi* (Old Norse) 'lump-fish'; though if so the attenuated *-fjord* ending found in other Norse Skye loch-names, like Snizort, has gone completely.

Sleat (Highland) 'Slope', from Gaelic *sliàbh* 'slope' (plural *sléibhte*) suits the topography best. 'Level place', from Old Norse *sletta,* 'level area', has also been proposed. Found as Slate, c. 1400.

Sligachan (Highland) 'Shelly place'. *Sligeach* (Gaelic) 'abounding in shells', *-an* (Gaelic diminutive termination).

Smoo Cave (Highland) 'The hiding place'. *Smuga* (Old Norse) 'hiding place'.

Snizort, Loch (Highland) Probably from *Sneisfjordr* (Old Norse) 'split firth'; but *Sneasfjordr* 'snow firth', has also been suggested. Noted in 1501 as Snesfurd. The form of the inner loch, divided by the Aird, supports the former explanation.

Soay (Western Isles) 'Sheep island'. *Sautha* (Old Norse) 'sheep', *ey* 'island'. There are numerous Soays. See also Fair Isle.

Solway Firth (Dumfries & Galloway) 'Firth of the muddy ford'. *Sol* (Old Norse) 'mud', *vath* 'ford', *fjordr* 'fiord', 'firth'. Recorded as Sulewad in a document of 1229.

Sorbie (Dumfries & Galloway) 'Bog settlement'. *Saur* (Old Norse) 'bog', 'mud', *by* 'settlement'. Noted on Blaeu's map, 1645, as Soirbuy. Soroba, by Oban, has the same source.

Spey, River (Highland; Moray) 'Hawthorn river' has been suggested, from early Gaelic *spiath*, cognate with *yspyddad* (Brittonic) 'hawthorn'; as has a link with the pre-Celtic root form **squeas*, 'vomit', 'gush'. On Ptolemy's map of c. CE 150 it appears as Tvesis; the form Spey is found in 1451. The Spean, which rises close to the Spey, is seen as a diminutive form of the same name, with the *-an* (Gaelic) diminutive ending.

Spittal (Moray; Highland) 'Refuge'. *Spideal* (Gaelic) 'refuge', 'hospice'. Spittals tend to be on remote hill passes, like Spittal of Glen Shee.

Sprouston (Borders) 'Sprow's farmstead'. *Sprow* (Old English personal name), *tūn* 'homestead', leading to Scots *toun* or *ton*. Noted as Sprostana, 1124.

Spynie (Moray) 'Hawthorn place', from a conjectured *spiath* (Gaelic) 'thorn', cognate with *yspyddad* (Brittonic) 'hawthorn' (*see* Spey). *Spionan* is still Gaelic for 'gooseberry'. Recorded as Spyny, c. 1220.

Stack Polly (Highland) 'Mountain of the river Pollaidh'. *Stac* (Gaelic) 'steep rock', *Pollaidh* 'pools', 'holes'.

Staffa (Argyll & Bute) 'Pillar island'. *Stafr* (Old Norse) 'staff', 'pillar', *ey* 'island'. The name reflects its vertical columns of basaltic rock. Staffin in Skye has the same derivation.

Start Point (Orkney) This headland is named from Old Norse *stertr*, 'tail'. Start Point in South Devon at the other end of Britain is from the cognate Old English *steort*.

Stenness (Orkney) 'Headland of stones'. *Stein* (Old Norse) 'stone', *nes* 'headland'.

Stepps (North Lanarkshire) 'Wooden road'. *Stap, stepp* (Scots) 'stave'. The reference is to a roadway made with wooden staves laid parallel, sometimes called a 'corduroy road'.

Stewartry (Dumfries & Galloway) The term records the former judicial stewardship by the earls of Douglas over the 'stewartry' or stewardship, of Kirkcudbright. See Mearns.

Stirling The name has been traced to a possible early Gaelic *srib-linn* 'stream pool', though other derivations have been proposed. A river name is most likely. It was recorded as Strivlin 1124, Estriuelin c. 1250, Striviling 1445, Sterling 1470.

Stobo (Borders) 'Hollow of stumps'. *Stub* (Old English) 'stump', *how* (Old English giving Scots howe) 'hollow'. Noted in the 12th century as Stoboc.

Stockbridge (Edinburgh) 'Bridge of tree-trunks', from Old English *stocc* 'tree-trunk', *brycg* 'bridge'. Stockbriggs in Lanarkshire has the same derivation.

Stonehaven (Aberdeenshire) Possibly 'stony landing place'. *Stan* (Old English) 'stone', *hyth* 'landing place'. Recorded in documents as Stanehyve 1587, Steanhyve 1629. Less likely is Old Norse *steinn* 'stone', and *hofn* 'harbour'.

Stornoway (Western Isles) Possibly 'Steering Bay'. *Stjorn* (Old Norse) 'steering', *vágr* 'bay'. The exact sense of this remains uncertain, and other possibilities include 'star bay' from Old Norse *stjorna*, 'star'. Recorded as Stornochway in 1511.

Stow (Borders) 'Place'. *Stow* (Old English) 'place', 'town'. It is unusual to find it on its own with no defining word.

Strachur (Argyll & Bute) 'Twisting valley'. *Srath* (Gaelic) 'wide valley', *cor* 'twist', 'bend'. Recorded as Strachore, 1368.

Straiton (South Ayrshire; Midlothian) 'Place on the Roman road'. *Straet* (Old English from Latin *strata*) 'road', *tūn* 'farmstead', 'settlement'.

Stranraer (Dumfries & Galloway) 'Place of the broad peninsula'. *Sron* (Gaelic) 'nose', 'peninsula', *reamhar* 'fat', 'thick'. This appears to be a reference to the location at the foot of the wider of the two arms of the Rinns of Galloway.

Strathaven (South Lanarkshire) *Strayven*. 'Wide valley of the Avon'. *Srath* (Gaelic) 'wide valley', *abhainn* 'river'. Noted as Straithawane, 1552. Here the terminal -th has been lost in speech but preserved in the spelling.

Strathbungo (Glasgow) 'Mungo's strath'. *Srath* (Gaelic) 'wide valley', *Mhungo* (Gaelic nickname) 'Mungo's'. It was the pet-name, meaning 'dear one', for St Kentigern.

Strathkinness (Fife) 'Strath of the water-head'. *Srath* (Gaelic) 'wide valley', *cinn* 'at the head of', *eas* 'water'. Noted as Stradkines, 1144.

Strathmiglo (Fife) 'Strath of the Miglo burn'. *Srath* (Gaelic) 'wide valley'; *mig* (Pictish) 'boggy area', *lo* Pictish suffix of uncertain form and meaning.

Strathmore (Perth & Kinross; Angus) 'The great wide valley'. *Srath* (Gaelic) 'wide valley'; *mór* (Gaelic) 'big', 'great'.

Strathpeffer (Highland) 'Wide valley of the shining stream'. *Srath* (Gaelic) 'wide valley'; *pevr* (Pictish-Brittonic) 'radiant one'. Noted as Strathpeffir, 1350.

Strathy (Highland) 'Of the strath'. *Srath* (Gaelic) 'wide valley', with -*ach* ending denoting place.

Strathyre (Stirling) Probably derived like Strachur, as 'twisty strath.'

Strawfrank (South Lanarkshire) 'Valley of the French'. *Srath* (Gaelic) 'wide valley, *frangaich* 'of the French', a reference to incoming landholders in medieval times. Noted as Strafrank in 1528.

Stroma (Highland) 'Island in the current', from Old Norse *straumr* 'current', 'stream', and *ey* 'island'. Recorded in 1150 as Straumsey.

Strome Ferry (Highland) 'Ferry of the channel'. *Straumr* (Old Norse) 'current', 'stream'.

Stromness (Orkney) 'The headland of the current'. *Straumr* (Old Norse) 'sea current, *nes* 'headland'. Records show it as Straumness in 1150. An earlier alternative name was Hamnavoe, meaning 'harbour on the bay'.

Stronachlachar (Stirling) 'The mason's point'. *Srón* (Gaelic) 'point', 'nose', *a* 'of', *chlachair* 'the mason's'.

Strone (Argyll & Bute; Stirling; Highland) 'nose', 'point', from *srón* (Gaelic) 'nose'. Strone by Dunoon is noted in 1240 as Strohon. Stron- or Sron- are often found as prefixes in other names.

Stronsay (Orkney) 'Star island', from Old Norse *stjorna*, 'star', and *ey*, 'island'. Noted as Stiornsay, 1150.

Strontian (Highland) 'Promontory of the beacon'. *Srón* (Gaelic) 'promontory', *teine* 'beacon'.

Succoth (Aberdeenshire; Argyll & Bute; other areas) 'The snout', meaning a point of ground between two converging streams, from Gaelic *socach*, 'snout'. The Biblical link seems coincidental.

Sullom Voe (Shetland) 'The gannets' fiord'. *Sulan* (Old Norse) 'gannets', giving Scots 'solan', *vágr* 'bay', 'sea inlet'.

Struan (Perth & Kinross; Highland) 'Stream place'. Gaelic *Sruthan* 'stream', 'burn'.

Struy (Highland; Perth & Kinross) 'river place'. Gaelic *sruth* 'river', 'stream', with *-ach* suffix denoting place; *sruthaigh* is its locative form. Often spelled as Struie.

Suie (Clackmannan; Aberdeenshire; Highland) 'Seat', from Gaelic *suidhe* 'seat', 'resting-place', often with the sense of having been a holy person's seat on a hill.

Suilven (Highland) Perhaps 'sun mountain'. *Sul* (Old Gaelic) 'sun', *bheinn* 'mountain'. However, the Gaelic name of Suilven is Beinn Buidhe (Gaelic) 'yellow'; its steep face looks directly to the setting sun. The derivation of the first element from *súlr* (Old Norse) 'pillar' has also been suggested.

Sumburgh (Shetland) Probably 'Sweyn's stronghold'. *Sweyn* (Old Norse personal name, *borgr* 'fort', 'stronghold'. Recorded as Swynbrocht in 1506.

Summer Isles (Highland) So called because they were used for summer grazing: a direct translation of Gaelic Na h-Eileanan Samhraidh.

Sunart, Loch (Highland) 'Sweyn's fjord'. *Sweyn* (Old Norse personal name), *fjordr* 'sea inlet,' 'firth'. Recorded as Swynwort, 1372.

Sutherland (Highland) 'Southern territory'. *Suthr* (Old Norse) 'south', *land* 'territory'. Recorded as Suthernelande, c. 1250. The Gaelic name is *Cataibh*, from the Pictish name *Cat*.

Sutors of Cromarty (Highland) The two headlands of the Cromarty Firth are *na Sùdraichean*, 'the tanners', in Gaelic, and the association of sound and meaning has produced the Scots form *Sutors*, 'cobblers'; recorded as the Sowteris, 1593. There is a Souter Head just south of Aberdeen, again adjacent to a Nigg Bay.

Symington (South Lanarkshire) 'Simon's farm'. *Symon* (Old English personal name), *tūn* 'farmstead'. The Simon was Simon Lockhart; the name noted around 1179, as Villa Symonis Lockard.

T

Tain (Highland) 'Water'. The name is taken from the river, Gaelic *tain*, ascribed to a pre-Celtic root form indicating 'river' or 'water'. Related river names are Teign in Devon, and Tyne in Northumberland and Lothian. It is recorded as Tene 1226, Thane 1483. The traditional Gaelic name is Baile Dhubhthaich, 'Duthac's Town', from the church of St Duthac here.

Tankerness (Orkney) 'Tancred's cape'. *Tancred* (Old Norse and Norman personal name), *nes* 'headland', 'cape'.

Tantallon (East Lothian) 'High-fronted fort'. *Din* (Brittonic) 'fort', *talgan* 'of the high front, or brow'.

Tarbat Ness (Highland) 'Cape of the isthmus'. *Tairbeart* (Gaelic) 'isthmus', 'portage point'; *nes* (Old Norse) 'headland'. Noted c. 1226 as Arterbert, with Gaelic *àrd*, 'high'. The Gaelic name is Rubha Thairbeirt.

Tarbert (Argyll & Bute; Western Isles) 'Place of the isthmus'. *Tairbeart* (Gaelic) 'isthmus, portage point'. Tarbet by Loch Lomond is recorded as Tarbart, 1392. Tarbert on Loch Fyne is Tairpirt Boetter, 'facing Bute', 711.

Tarff (Highland; Perth & Kinross; Dumfries & Galloway) 'Bull stream', 'bull place'. *Tarbh* (Gaelic) 'bull'. This frequently found name also takes the form Tarves and Tarvie.

Tarland (Aberdeenshire) 'Bull's enclosure', from *tarbh* (Gaelic) 'bull', and *lann* 'field, enclosure'. Noted as Tarualund, 1183.

Tarradale (Highland) 'Bull's dale'. From Old Norse *tarfr*, 'bull', and *dalr*, 'dale'.

Tarskavaig (Highland) Suggested as 'Cod bay'. *Thorskr* (Old Norse) 'cod', *vaig* (from *vik*) 'bay'. Alternatively, from *Tar* (Gaelic prefix) 'across from'; meaning 'across from Scavaig'. This sense may have been 'grafted' on to the Norse name after the decline of Norse speech in Skye. *See* Scavaig.

Tay, River, Strath, Loch, Firth (Perth & Kinross) The river is noted in the 1st century CE by Tacitus as Taus, and by Ptolemy, around AD 150, as Tava. Its name may derive from a conjectured Brittonic **tausa*, meaning 'silent one' or 'strong one' – aspects of the river's controlling deity – or simply 'flowing' (see also Teith).

Taynuilt (Argyll & Bute) 'House by the stream'. *Taigh* (Gaelic) 'house' *an* 'of the', *-uillt* 'of the stream'.

Tayport (Fife) The current name dates only from 1888; before then it was successively called Scotscraig, South Ferry, Portincraig, Ferryport-on-Craig and South Craig, all names referring either to its ferry across the Tay or to the crag on which the town is situated.

Tayvallich (Argyll & Bute) 'House in the pass'. *Taigh* (Gaelic) 'house', *bhealaich* 'of the hill pass'.

Teith, River (Stirling) Another ancient river-name whose origins are unclear, but which is presumed to stem from the same pre-Celtic, or possibly non-Indo-European, root element **tau*, 'flowing, melting', as Tay, Teviot.

Templand (Aberdeenshire; Dumfries & Galloway) 'Temple-land', indicating land once belonging to the Knights Templar; the placename Temple (Lothian and elsewhere) has the same sense. Gaelic *teampull* 'temple', 'church' names in the Hebrides simply indicate a church.

Teviot, River (Borders) As with so many river-names, its origin goes far back into unrecorded history; it may be linked to the pre-Celtic root form *tau*, 'flowing, melting' (*see* Teith). Noted c. 600 as Teiwi, c. 1160 Teuiot.

Threipland (Aberdeenshire; South Lanarkshire) 'Debatable land'. *Threap* (Middle English and Scots) 'scold', 'dispute'; and 'land'. A tract at one time under disputed ownership.

Thundergay (North Ayrshire) 'Backside to the wind', from Gaelic *tòn* 'backside', *ri gaoith*, 'to the wind', presumably a comment on the Arran site's exposure. There is also Tunregaith in South Ayrshire, and Timothy Pont's 1645 map shows a Tonreghe near Whithorn.

Thurso (Highland) 'Bull's river'. *Thjor-s* (Old Norse) 'bull's'; *aa* 'river'. Recorded as Thorsa in a document of 1152, and this still accords with local pronunciation. The origin may be pre-Norse, since Ptolemy's name for the neighbouring headland, Tarvedum (c. AD 150) suggests a conjectured Pictish *tarvo-dubron*, 'bull-water'. Later Norse speakers perhaps made a false relation to the god-name Thor.

Tighnabruaich (Argyll & Bute) 'House of the bank'. *Taigh* (Gaelic) 'house', *na* 'of the', *bruaich* 'bank'.

Tillicoultry (Clackmannanshire) 'Hill-slope in the back land'. *Tulach* (Gaelic) 'hill', 'hill-slope', *cul* 'back'; *tìr* 'land'. Recorded as Tulycultri, 1195.

Tillienaught (Aberdeenshire) 'Bare hill'. *Tulach* (Gaelic) 'hill', 'hill slope', *nochd* 'bare'.

Tinto (South Lanarkshire) 'Beacon hill'. *Teine* (Gaelic) 'fire', 'beacon', *ach* (suffix denoting place). Noted as Tintou, c. 1315, but into the 19th century it was known as Tintock.

Tiree (Argyll & Bute) Possibly 'land of corn'. *Tìr* (Gaelic) 'land', *eadha* 'corn'. The Old Irish personal name *Ith* has also been suggested as the source of the latter part, giving 'Ith's land'. Early forms include Tir Iath, 6th century, Terra Ethica, c. 700, Tiryad, 1343.

Tobermory (Argyll & Bute) 'Mary's well'. *Tobar* (Gaelic) 'well', *Moire* 'Mary's'. Noted as Tibbirmore, 1540.

Tomatin (Highland) 'Juniper hill'. *Tom* (Gaelic) 'hill', 'knoll', *aitionn* 'juniper'.

Tomintoul (Moray) 'Little hill of the barn'. *Tom* (Gaelic) 'hill', *an t-sabhail* 'of the barn'.

Tongland (Dumfries & Galloway) 'Tongue-shaped piece of land', from Old English *tunge*, 'tongue', cognate with Old Norse *tunga*; and Old English *land*. Noted as Tuncgeland, c. 1150.

Tongue (Highland) 'Tongue or spit of land'. *Tunga* (Old Norse) 'tongue'.

Tore (Highland) 'Bleaching place'. The name has sometimes been construed as Gaelic *torr*, 'hill', but the Gaelic name is *an Todhar*, 'the bleaching place'. See Balintore.

Torness (East Lothian; Highland; Shetland) In Lothian, 'mound of the headland': *Torr* (Gaelic) 'mound', 'hill'; *nes* (Old Norse) 'headland'. Torness near Inverness may be *Torr nan eas* (Gaelic) 'mound of the stream or waterfall'.

Torphins (Aberdeenshire) 'White mount'. *Torr* (Gaelic) 'mound', 'hill', *fionn* 'white', 'fair'. Its terminal -*s* may indicate a plural in the original Gaelic name. Torphin in south Edinburgh shows a singular form, with the same derivation.

Torridon (Highland) The source is not confirmed. The first part element be *torr* (Gaelic) 'hill(s)', but a meaning based on the Irish Gaelic verb *tairbhert*, 'transfer' has also been suggested, on the supposition that Glen Torridon was a portage route from the head of Upper Loch Torridon to Loch Maree. In 1464 it was recorded as Torvirtane.

Torry (Aberdeen) 'Hilly place'. *Torr* (Gaelic) 'hill', *aidh* (suffix denoting place).

Tough (Aberdeenshire) 'Hills' or 'hilly place'. *Tulach* (Gaelic) 'hill, ridge'. Recorded as Tulluch, c. 1550, Towch 1605. The Touch Hills to the south of Flanders Moss, recorded 1329 as Tulch, have the same derivation.

Tradeston (Glasgow) This Glasgow district was developed as a residential area around 1790 by the Glasgow Trades House, a guild of merchants.

Tranent (East Lothian) 'Village by the valley'. *Tref* (Brittonic) 'settlement', *yr neint* 'by the valley'. The form Trauernent is found c. 1127.

Traquair (Borders) 'Homestead on the river Quair'. *Tref* (Brittonic) 'homestead', the river-name is probably from Celtic **vedra* 'clear one'; as with Weir. Older forms include Treverquyrd in 1124.

Trearne (Renfrewshire) 'House among the sloes', from Brittonic *tref* 'house', and *àirne*, 'the sloe trees'. One of the rare unaltered Tre- prefixed names in Scotland; but Old English *trow-aerne* 'timber house' has also been suggested.

Trool, Loch and **Glen** (Dumfries & Galloway) 'Loch of the stream'. *Loch* (Gaelic) 'lake', 'loch'; *an t-* 'of the', *sruthail* 'stream'.

Troon (South Ayrshire) 'Headland'. *Trwyn* (Brittonic) 'headland', 'point'. A less likely derivation is *an t-sron* (Gaelic) 'nose', 'point'. Recorded as le Trone in 1371 and known as The Troon into the 19th century. The same root name is found in Duntrune, Argyll, 'fort on the headland'.

Trossachs, The (Stirling) Apparently 'the cross-hills', from a Brittonic word cognate with Welsh *trawsfynnydd* 'cross-hill', rendered into Gaelic form in modern times as Na Trosaichean. *Tros* (Old Welsh) signifies 'across'.

Trotternish (Highland) 'Thrond's Headland'. *Throndar* (Old Norse personal name), *nes* 'headland', recorded as Trouternish, 1309, Tronternesse in the mid 16 century. With Minginish and Vaternish, one of the three main divisions of the Isle of Skye.

Truim, River and **Glen** (Highland) 'Of the elder trees'. *Trom* (Gaelic) 'elder tree'.

Tullibardine (Perth & Kinross) 'Hill of warning'. *Tulach* (Gaelic) 'hill', *bàrdainn* 'warning'. The reference is to a signal beacon. Noted as Tulibarden, 1234.

Tullibody (Clackmannanshire) Perhaps 'Yellow Hill'. *Tulach* (Gaelic) 'hill', *boidhe*, a form of *buidhe* 'yellow'. Some older forms had the prefix *Dun-*, perhaps through Brittonic influence: Dunbodeuin, 1147.

Tullybelton (Perth & Kinross) 'Beltane hill'. *Tulach* (Gaelic) 'hill', 'hill slope', *Bealtainn* 'Beltane', the Celtic May feast, when ritual fires were lit on conspicuous hilltops.

Tullynessle (Aberdeenshire) 'Hill of spells' has been proposed. *Tulach* (Gaelic) 'hill', *an* 'of', with *eoisle* by metathesis from *eólas* 'knowledge', 'spell'. Recorded as Tulynestyn, c. 1300. Or possibly from a lost Pictish word cognate with Welsh *iselfynydd*, 'low hill'. Esslemont in the same region appears to reverse the same elements.

Tummel, River, Strath and **Loch** (Highland) 'Dark (river)'. *Teimheil* (Gaelic) 'dark'. Like many other river-names, it may be older than this suggests, from a pre-Celtic form that implies the same meaning.

Tundergarth (Dumfries & Galloway) Though the suffix may be Gaelic *gart*, 'enclosure', the early 13th-century form of Thonergayth also suggests the same derivation as Thundergay.

Turnhouse (Edinburgh) Perhaps 'hill of the spectre', from *torr* (Gaelic) 'hill', *na* 'of', *fhuathais* 'of the spectre'.

Turret, River and **Glen** (Highland; Perth & Kinross) 'Little dry stream', from Gaelic *tur*, 'dry', and *that*, a suffix indicating 'small'. The reference is to a stream that dries out in summer.

Turriff (Aberdeenshire) Possibly 'hill of anguish'. *Torr* (Gaelic) 'hill', *bruid* 'anguish' or 'a stab'. An exact derivation remains uncertain. Records show Turbruad, in the *Book of Deer* c. 1000, Turrech 1300, Turreff 1500.

Tuskerbuster (Orkney) 'Peat-cutter's farm'. *Torf* (Old Norse) 'peat', *skeri* 'cutter', *bolstadr* 'farmstead'.

Twatt (Orkney) 'Clearing', 'settlement'. *Thveit* (Old Norse) 'clearing', 'meadow', cognate with Old English 'thwaite'.

Tweed, River (Borders) A river-name of uncertain derivation. It may stem from the same Celtic root form **tau* or **teu*, indicating 'strong, silent', or 'flowing', ascribed to Tay and Tyne. This in turn has been linked to Sanskrit *tavas*, meaning 'surging' or 'powerful'. Recorded around 700 as Tuuide.

Tyndrum (Argyll & Bute) 'House on the ridge'. *Taigh* (Gaelic) 'house', *an* 'on the', *druim* 'ridge'.

Tyne, River (East Lothian) This river-name stems probably from a pre-Celtic root indicating 'water', cognate with Tain and Tay.

Tyninghame (East Lothian) 'Village of the dwellers by the Tyne'. *Tyn* (see Tyne); *inga* (Old English) 'of the people', *ham* 'settlement', 'village'.

U

Uamh, Loch nan (Highland) 'Loch of the Cave'. *Loch* (Gaelic) 'lake, loch', *nan* 'of', *uamh* 'cave'.

Uddingston (South Lanarkshire) 'Oda's people's farmstead'. *Oda* (Old English personal name), *inga* 'of the people', *tūn* 'farmstead'. An early form of 1296 is recorded as Odistoun, just 'Oda's farmstead'. There is also Uddington in the same district.

Udny (Aberdeenshire) 'Streams'. *Alltan* (Gaelic) 'streams', with -*ait*, suffix denoting place. Around 1400 it was recorded as Uldeny.

Ugie, River (Aberdeenshire) 'Stream of nooks and corners', from *ùigeach* (Gaelic) 'nook', 'hollow'.

Uig (Highland; Argyll & Bute; Western Isles) 'Bay'. *Uig*, a Gaelic form of *vik* (Old Norse) 'bay'.

Uist, North and **South** (Western Isles) 'An abode'. *I-vist* (Old Norse) 'in-dwelling'. It was recorded in 1282 as Iuist and in the 14th century as Ywest. The modern Gaelic form is Uibhist, and the meaning appears to correspond with that suggested for Lewis.

Ulbster (Highland) 'Ulf's farm', from Old Norse personal name *Ulfa* (signifying wolf-like), and *bolstadr* 'farmstead'. The suffix is a compressed form of the -bister names of Orkney, as with numerous other northern mainland places. Recorded as Ulbister, 1538.

Ullapool (Highland) 'Olaf's settlement'. *Olaf* (Old Norse personal name), *bol* (Old Norse mutated form of *bolstadr*) 'farmstead', 'settlement'. Recorded in 1610 as Ullabill.

Ulva (Argyll & Bute) 'Ulf's island'. *Ulfa* (Old Norse personal name or nickname) 'Wolf', *ey* 'island'. Recorded in 1473 as Ulway.

Unapool (Highland) 'Uni's farm'. *Uni* (Old Norse personal name), *bol* from *bolstadr* 'farmstead', 'settlement'. See Eriboll.

Unst (Shetland) Possibly 'Eagles' nest'. *Orn* (Old Norse) 'eagle', *nyst* 'nest'. Recorded in a document of around 1200 as Ornyst. But possibly a lost Pictish name, see Fetlar.

Urie (Aberdeenshire) Perhaps 'place of the yews'. *Iubharach* (Gaelic) 'of yews', though a derivation from *uar* 'landslip', 'water-spout', may also be possible, with the often-found river-name suffix *-aidh*. A further suggestion is *uidhre*, genitive of *odhar* 'brown'. In Glen Ury, by Stonehaven, an alternative spelling is found.

Urquhart (Highland) 'Woodside'. *Air* (Brittonic–Pictish) 'on', 'upon', *cardden* (Pictish) 'thicket', 'wood'. Adamnan, c. 700, records Aírchardan; the Urquhart form is found from 1340. In Gaelic it is Urchardainn.

Urr, River (Dumfries & Galloway) Another river-name from remote antiquity, ascribed to a pre-Celtic origin. It has been compared to Basque *ur* 'water', but it would be rash to draw conclusions from this. Recorded c. 1280 as Urrer.

Urrard (Perth & Kinross) 'Fore-headland', from Gaelic *air* 'on', 'upon'; *àirde* 'height'.

Urray (Highland) Perhaps 'remade fort'. *Air* (Gaelic) 'on', *rath* 'ring fort', with the notion of a fortification on top of an earlier one. Noted as Vrray, 1546.

V

Vatersay (Western Isles) Perhaps 'Glove island'. *Vottr-s* (Old Norse) 'glove's', *ey* 'island'. Recorded as Vatersa, 1580. The Gaelic form is Bhatarsaigh.

Venue, Ben (Stirling) 'Little mountain'. *Beinn* (Gaelic) 'mountain', *mheanbh* 'small'. Noted as Benivenow, 1794.

Vennachar, Loch (Stirling) 'Horned loch'. Although 'Loch of the fair corrie', from *loch* (Gaelic) 'lake', 'loch', *bhana* 'fair', *choire* 'of the mountain corrie', has been suggested, an older form of the name is Banquhar, c. 1375, and the Gaelic form is Loch Bheannchair, making it cognate with Banchory, from *beannchar*, horn-shaped.

Voil, Loch (Stirling) Perhaps 'lively', from *beò* (Gaelic) 'life', 'breath'. The Gaelic name is Loch Bheothail.

Vorlich, Ben (Stirling, Perth & Kinross) 'Mountain of the sea-bag'. *Beinn* (Gaelic) 'mountain', *muir* 'sea', *bolc* 'bag'. The reference to a bag-like bay in the adjacent loch has been accepted in the case of Ben Vorlich (Loch Lomond), noted on Blaeu's map in 1645 as Benvouirlyg, but disputed in that of Loch Earn, which has also been linked to a hypothetical Old Gaelic personal name, *Murlag*.

Vrackie, Ben (Perth & Kinross) 'Speckled mountain'. *Beinn* (Gaelic) 'mountain', *bhreachaidh* 'speckled'. Ben Bhraggie, above Golspie in Sutherland, has the same derivation.

W

Walkerburn (Borders) 'Waulker's stream'. *Waulker* (Scots, from Old English *walcere*) 'fuller of cloth', *burna* (Old English) 'stream', giving Scots 'burn'.

Walls (Orkney; Shetland) 'Bays', from Old Norse *vágar* 'bays'. Walls at the south of Hoy is noted as Vagaland in the *Orkneyinga Saga*, c. 1225.

Wallyford (Midlothian) 'River ford'. Old English *waelle* 'spring', 'stream', and Scots 'ford'. Noted as Walford, 12th century. No link to Scots *wally* 'fine'.

Wamphray (Dumfries & Galloway) Perhaps 'cave of the offerings'. *Uamh* (Gaelic) 'cave', *aifrionn* 'place of offerings', 'chapel'. Recorded as Vamphray, 1275.

Wanlockhead (Dumfries & Galloway) Stream-head of the 'white flat stone'. *Gwyn* (Brittonic) 'white', *llech* 'flat stone'. The name may have first denoted the Wanlock Water, with the English *-head* suffix added at a much later date. Recorded as Wenlec, 1563.

Ward Hill (Orkney; Shetland) 'Look-out hill'. Old Norse *vardr* 'watch', 'guard'. There are numerous Ward Hills in Shetland especially, sometimes spelled Vord.

Wardlaw (Highland; other regions) 'Look-out Hilll', from Old English *weard*, 'guard', 'watch', and *hlaew*, 'hill', becoming Scots 'law'. Noted 1210 as Wardelaue. See Fare, Hill of.

Waterbeck (Dumfries & Galloway) This appears to be a rare example in Scotland of 'beck' from Old Norse *bekkr* 'stream', with *vatn* 'water'.

Waterloo A number of locations have this name in commemoration of the battle of Waterloo in 1815.

Waternish/Vaternish (Highland) 'Water headland'. *Vatn* (Old Norse) 'water', *nes* 'headland'. Noted as Watternes, 1501. With Minginish and Trotternish, one of the three main divisions of the island of Skye.

Watten (Highland) 'Water', 'lake'. *Vatn* (Old Norse) 'water'. The village takes its name from the loch, now inevitably known as Loch Watten. Noted as Watne, 1230.

Wauchope (Dumfries & Galloway) 'Den of strangers'. *Walc* (Old English) 'stranger', 'foreigner', *hop* 'enclosed valley'. Noted as Walchope, 1214.

Weem (Perth & Kinross) 'Cave'. *Uamh* (Gaelic) 'cave'.

Weir, River (Renfrewshire) The derivation goes back to Celtic **vedra* 'clear one', via Brittonic *gueir* 'clear', 'white'.

Wemyss (Fife; North Ayrshire) 'Caves'. *Uamh* (Gaelic) 'cave'. East Wemyss and West Wemyss are noted as Wemys 1239. Port Wemyss on Islay is named for a 19th century landlord.

Westray (Orkney) 'West island'. *Vestr* (Old Norse) 'west', *ey* 'island'; found as Westray in the *Orkneyinga Saga*, c. 1225.

Whalsay (Shetland) 'Whale's island'. *Hval-s* (Old Norse) 'whale's', *ey* 'island'. Nted as Hvalsey in the *Orkneyinga Saga*, c. 1225.

Whitburn (West Lothian) 'White stream'. *Hwit* (Old English) 'white', *burna* 'stream'. This Anglian form gave Scots 'burn', which continued in use here whereas 'brook' largely replaced it in England. Recorded in 1296 as Whiteburne.

Whithorn (Dumfries & Galloway) 'White house'. *Hwit* (Old English) 'white', *erne* 'house'. Called Candida Casa (Latin 'white house') on its foundation in AD 397 by St Ninian, and noted as Hwitan Aerne c. 890.

Wick (Highland) 'Bay'. *Vik* (Old Norse) 'bay'. It was recorded as Vik in 1140, Weke in 1455. Its Gaelic name is *Inbhir-Uig*, 'river-mouth bay'.

Wigtown (Dumfries & Galloway) 'Wicga's farm'. *Wicga* (Old English personal name), *tūn* 'farmstead'. Recorded as Wigeton in 1266.

Winchburgh (West Lothian) 'Winca's fort'. *Winca* (Old English personal name), *burh* 'fortified place'. Recorded as Wynchburch, 1375.

Windygates (Fife) 'Windy gap'. *Geat* (Old English) 'gate' became Scots *yett*, also with the meaning of hill pass or gap, as in Yetts of Muckhart. Although an old form Windeyetts is recorded, the *g*- form has been preserved.

Wishaw (North Lanarkshire) Probably 'Willow wood'. *Withig* (Old English) 'willow', *sceaga* 'wood' giving Scots 'shaw'.

Wormit (Fife) 'Wormwood'. *Wormit* is the Scots word for 'wormwood', and here presumably refers to a plantation of trees. Recorded as Wormet, 1440.

Wrath, Cape (Highland) 'Turning point'. Cape (from Latin *caput*, 'head', via Old French *cap*) 'headland'; *hverfa* (Old Norse) 'to turn'. The only 'Cape' name, recorded as Wraith, 1583. The landward district, the Parph, preserves a Gaelic form of the Norse name.

Wyvis, Ben (Highland) 'Majestic mountain'. *Beinn* (Gaelic) 'mountain'; *Uais* 'noble', 'majestic', a shortened form of *uasal*, 'proud'. *Fhuathais* 'of the bogle' or 'goblin', has also been suggested, but the first meaning is much more probable. It was noted as Weyes in 1608.

Y

Yarrow, River (Borders) 'Rough (river)', from *garw* (Brittonic) 'rough'. Recorded as Gierua, c. 1120.

Yell (Shetland) 'Barren place'. *Geldr* (Old Norse) 'barren'. Noted in the *Orkneyinga Saga*, c. 1225, as Ala; and in later documents as Jala, Jella and Yella.

Yester (East Lothian) 'House', 'dwelling'. *Ystre* (Brittonic) 'dwelling'.

Yetholm (Borders) 'Village of the pass'. *Geat* (Old English) 'gate', 'gap', become Scots yett, with *ham* 'village'. Old Norse *holmr* 'island', including 'river island', would seem to be the source of the suffix, but early records show the *ham* form: Gatha'n c. 800, Jetham 1233, Kirkyethame c. 1420.

Yoker (Glasgow) 'Low ground'. *Iochdar* (Gaelic) 'low-lying ground'. Found in this form 1505.

Ythan, River (Aberdeenshire) 'Talking stream'. *Iaith* (Brittonic-Pictish) 'language', 'talk'; with the *-on* ending found in many stream-names, deriving from a hypothetical early Celtic form *-*ona*, indicating 'water'. Found in 1373 as Ethoyn, 1477 Ithane.